timber decking

the professionals' manual

2nd edition

Patrick J Hislop RIBA,
TRADA Technology

TRADA Technology Ltd
Chiltern House
Stocking Lane
Hughenden Valley
High Wycombe
Buckinghamshire HP14 4ND

tel: +44 (0)1494 569600
fax: +44 (0)1494 565487
email: information@trada.co.uk
website: www.trada.co.uk

2nd edition 2006

ISBN 1-900510-52-9 (10 digit)
978-1-900510-52-3 (13 digit)

1st edition published in 1999 by TRADA.

Whilst every effort is made to ensure the accuracy of the advice given, the company cannot accept liability for loss or damage arising from the information supplied.

Cover photographs: main, HLD Ltd
Inset photographs: HLD Ltd, 2nd left Ecotimber Ltd (supply of FSC certified hardwood and photo)

Printed on chlorine-free paper with a 50% recycled content

timber decking the professionals' manual

TRADA Technology is an independent consultancy company providing a wide range of commercial and training services to the timber and construction industries. Prior to 1994 it was wholly owned by TRADA, the Timber Research and Development Association. It is now a member of the TTL Chiltern Group of companies and is TRADA's appointed provider for its Research and Information Programmes and for the administration of its membership services.

TRADA Technology acknowledges the input and support from the Timber Decking Association in the revision of this manual.

The Timber Decking Association is an independent technical and advisory organisation formed in 1998 to establish and promote the materials, design and installation practices required to create high performance decks and associated outdoor structures in wood.

The TDA also operates DeckMark™, a third party quality assurance scheme for the materials and contractors involved in deck construction.

Timber Decking Association
5 Flemming Court
Castleford
West Yorkshire
WF10 5HW

tel: 01977 558147
email: info@tda.org.uk
website: www.tda.org.uk

Contents

1 Use of timber for decking

Figure 1.1 Multi-level deck with pool
Photo: HLD Ltd

The use of timber for external decking has a long history ranging from its use as medieval walkways in the fens, to the verandas, railway platforms, piers and jetties of the nineteenth century. In the major timber growing areas of the world, such as North America and Scandinavia, wood was used for sidewalks in the cities and lasted for many years before eventually being replaced by paving.

The deck however largely developed as an uncovered version of the verandas typical of frame houses of the tropics and the Southern states of America. They are an integral feature of open-air living which is such an aspect of the lifestyle of these areas. However, timber decks are now common all over the world, including the UK.

Figure 1.2 Deck at visitors centre
Photo: HLD Ltd

The popularity of outdoor living today is evident in the increased numbers of pergolas, conservatories, and summerhouses as well as external sitting areas such as decks and patios associated with UK housing. It is also reflected in the increased popularity of outdoor dining or drinking areas attached to pubs, restaurants and hotels where 'continental' habits of eating and drinking are being adopted by the British public. Many of these open-air spaces are in the form of timber decks because these are easy and quick to build, require less groundwork than masonry construction, particularly on sloping or rough sites, and can blend in well with existing landscape features. Timber decks are frequently associated with waterside developments where the decked area may extend partially over the water, providing a 'maritime' character.

Figure 1.3 Cliff top viewing deck
Photo: HLD Ltd

A significant expansion over recent years has been in the use of timber decking for large areas of public open spaces. The availability of durable hardwoods with full environmental certification has been a strong influence on this market. Further information is given in Appendix I.

Timber decks are particularly popular for sloping sites, where they can project out over land that would otherwise require major groundworks in order to provide a sufficiently level area for a conventional terrace. Often stepped in level to provide for various different activities, they frequently include pergolas, windbreaks, screens, benches, pools, spa pools, storage spaces and barbecues.

Figure 1.4 Massaranduba boardwalk at the National Assembly for Wales. Supplier of FSC certified timber and photos: Ecotimber Ltd

1.1 Advantages of using timber

Figure 1.5 High level decks
Photo: TDA/Peter Brook Design

A patio or terrace can be made from many materials – brick, stone or precast concrete slabs, ceramic tiles, metal or wood.

Timber has several practical advantages over other materials in that it can:

- Reduce excavation, foundations, and retaining walls – particularly if there are slopes or changes of level across the site.
- Avoid wet construction, other than post foundations, allowing simple and quick erection.
- Reduce heavy labour – digging foundations, mixing and placing of concrete, moving and laying masonry slabs etc.
- Provide for simple changes of level with ramps or steps and include other ancillary features such as handrails and balustrades.
- Allow for easy extension and addition of other features in the future such as seats, pergolas, planters, storage facilities etc.
- Provide simple ways of achieving complex or curvilinear shapes in order to fit more readily into existing spaces or landscape layout.
- Include the facility to add colour in order to blend in with, or enhance, the surroundings.
- Provide a surface on which it is comfortable to walk, sit or lie directly. This is particularly important if babies or toddlers are involved, as sliding, crawling or falling on wood is less damaging than on other hard surfaces.

Figure 1.6 Stepped deck
Photo: TDA

1.1.1 Ecological advantages

Besides the practical advantages of using timber for patios and terraces, it also has an additional advantage over other forms of construction in ecological terms. Wood is a renewable material, requiring little energy in conversion and production, and efficient use is made of the waste products, such as sawdust and chippings. Ultimately wood is easily recycled or disposed of, although care must be taken with treated material.

Figure 1.7 Yorkshire Dales Visitor Centre
Photo: TDA/Peter Brook Design

Softwood for decking and associated structures generally comes from the managed coniferous forests of North America, UK or Scandinavia. Temperate hardwoods, such as European oak, are obtained from managed forests in the British Isles or Europe. Other suitable hardwoods come from Australasia and the tropics and increasingly these too can be obtained from managed forests, certified by reputable national or international organisations.

Appendix I gives more infomation and includes the properties of some species of timber commonly used for decking.

Figure 1.8 Ipê decking outside Stratford Station, London

1.2 Perceived disadvantages

Many of the perceived disadvantages of using timber for decking arise from a lack of understanding of the material and a lack of attention to design detailing.

- Possible risk of decay
  This can be eliminated by correct detailing and the use of suitably durable species, or timber pressure treated with preservative, see Section 3.1.

- A high maintenance requirement
  The need for surface maintenance can be eliminated by using timbers that do not need a protective finish and can be allowed to bleach out naturally. However, where the species chosen requires some surface protection, or the aim is to either maintain the colour of the original wood, or to add some other colour, then either clear water-repellent seals, oils, or pigmented stains can be used. Clear finishes will not prevent the colour bleaching from the wood in the long term and will require quite frequent re-coating as there is no pigment to resist the effect of ultra-violet light. The need to maintain these finishes will depend partially on the choice of species and also on the degree of surface wear and exposure. Re-finishing with translucent or opaque stains is a relatively simple process involving brushing off surplus dirt and loose material and applying new coats of stain. See Chapter 8.

- Slipperiness and wear on the deck surface
  Timber decks can sometimes be considered 'slippery', but this is at least partially due to water that lies on the surface of wood. It is therefore important to ensure that any surface water runs off as quickly as possible and the surface is kept clean of dirt. If a deck is ramped to

provide access or includes steps, it may be advisable to improve the slip resistance by additional measures. See Section 3.6.

- Risk of fire
  Experience has shown that timber decks are very resistant to ignition. Even where there is a particular source of fire, such as a barbecue, it is very difficult to ignite the deck boards from above, although it is sensible to take suitable precautions such as ensuring that litter or vegetation cannot collect beneath the deck.

1.3 Functional requirements

The functional requirements of a deck will vary considerably depending on such factors as the level of pedestrian traffic, the degree of exposure, environmental conditions, or the particular activities the deck must accommodate, such as dining, sunbathing, or spectating. The aim of this manual is therefore to provide sufficient advice on the detail design and specification of decks appropriate to these various functions, as well as including basic information on the choice of species, preservative treatment and finishing.

1.4 Good practice and expertise

By establishing principles of good practice in design and specification, the aim is to ensure that timber decks, whether domestic or public, will not only be suitable for their purpose, but will also meet the desired service life, with the minimum of maintenance.

To achieve this, designers, contractors and clients need to understand the quality of design, materials and construction required in building decks. For this reason it is important to choose material suppliers and contractors who have experience in the supply of the materials for decking and their construction.

The Timber Decking Association (TDA) was set up by major materials suppliers and contractors in the decking industry to provide expertise to specifiers, builders and purchasers, see www.tda.org.uk.

Sponsored by the TDA, a demonstration deck was built at TRADA Technology's offices in Buckinghamshire in 2000. Figure 1.9 shows the construction of this deck.

Figure 1.9 above Construction of deck by the TDA at TRADA Technology's offices in 2000

right:The deck in 2006

2 Pre-planning

Before considering the materials and method of construction for a deck it is important to consider its size in relation to the space available and also the best location relative to the house or any other building. The size may be determined by the proximity to the boundaries such as roads or neighbouring properties, changes of level on site, natural features such as trees, access to both the building and other surroundings, and the cost. If funds are limited it is better to reduce the initial size of the deck rather than to skimp on material in the construction. With timber decks it is relatively easy to enlarge the deck later, possibly at a different level, to accommodate any fall on the site. Similarly, additional features such as pergolas or screens can be added as the need arises, even after the deck has been in use for some time. Money is better spent initially on achieving a robust and durable basic construction than on additional features.

Figure 2.1 Sunny deck, taking advantage of the view
Photo: Richard Burbidge

The location of the deck should take into account the following points, illustrated in Figure 2.4:

- Will the deck be in the sun when this is desirable, including low angle winter sun?
- Will there be some shade from buildings or trees in high summer?
- Will some areas be in permanent shadow due to shading from buildings or heavy foliage?
- Will the deck surface be permanently moist due to fallen leaves, overhanging foliage, or subject to regular wetting from swimming pools, ponds, fountains, etc?
- Will the area be adequately protected by buildings, walls, planting etc from prevailing winds? Will it be necessary to provide additional screens etc for protection from wind?
- Will the deck be at a location and height to provide the best view, taking into account the height of any guard rails and balustrades, which may in themselves form a visual barrier?
- Will building the patio or deck spoil any existing view from the property?
- Will any adjacent building offer some degree of protection from driving rain or snow and so prolong the life of any surface finishes?
- Will the design of the deck be safe for children at play, possibly when unsupervised?
- Will there be sufficient privacy from neighbouring buildings, roads, footpaths etc? Will the deck be overlooked from upper floor windows of other properties?
- Will the location of the deck restrict access to any existing services or will existing services, such as drains, cause difficulties in the construction of the deck?

Figure 2.2 Deck in partial shade
Photo: Richard Burbidge

Figure 2.3 Deck at Kew Gardens sited to provide a viewing platform
Photo: HLD Ltd

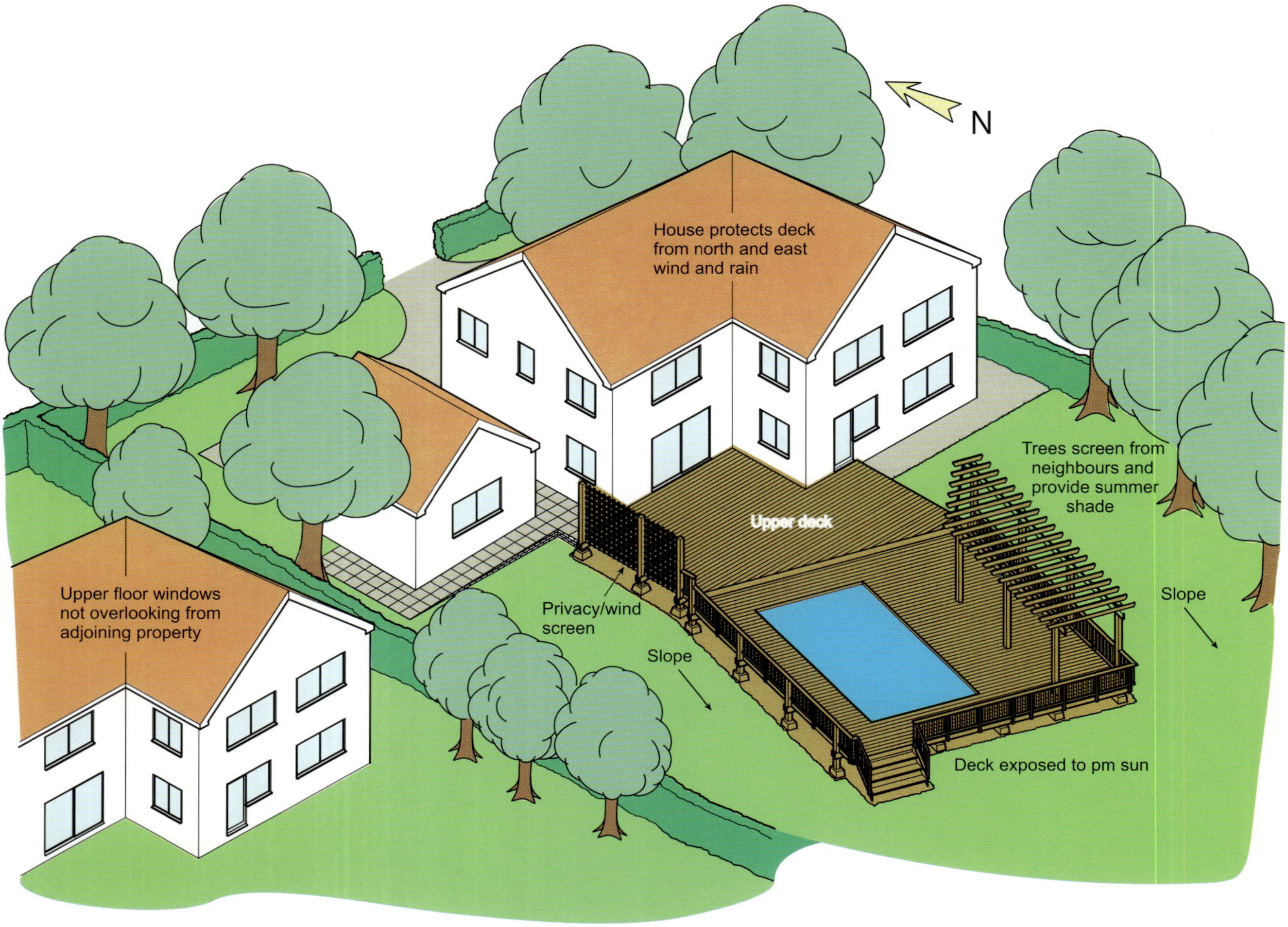

Figure 2.4 Planning a timber deck

2.1 Appearance considerations

The detailed decisions on the construction, layout, specification and finishes of a deck will have a significant effect on the finished appearance.

One of the first decisions to make is which way the boards should run relative to the shape of the area to be covered. Where this area is relatively square, there may be no practical advantage in running the boards one way or the other and the decision can be made largely on a visual preference. Looking along the length of the boards emphasises the joints and consequently visually 'extends' the deck's length. Running the boards across the line of sight reduces the emphasis on the joints, but tends to visually widen the space. If the space to be decked is long, and relatively narrow, there are practical advantages in running the boards across the short dimension. The primary advantage of this arrangement is that it is usually possible to use single lengths of board for the width whereas the boards would probably have to be butt jointed end to end if they run the length of the space. This can lead to a more complicated arrangement of the supports and fixing.

If the boards run across the short dimension of the deck any curves in plan can also be more easily taken up by varying the joint width between the boards, or, if necessary, introducing tapered boards to take up tighter bends. For angled bends, the boards can be arranged either way but it is somewhat easier to cut and fix boards laid lengthways rather than across the deck. For right-angled bends it is usually preferable to change the board direction on the corner in order to continue spanning the boards across the short dimension. Whichever pattern is chosen the direction of the boards will be visually emphasised by any grooving in the top surface, as well as by the joints between the boards. A popular alternative is to run the boards diagonally across the joists rather than at right angles. See Figure 2.5. Detailed consideration of deck board design and fixing is included in Chapter 5

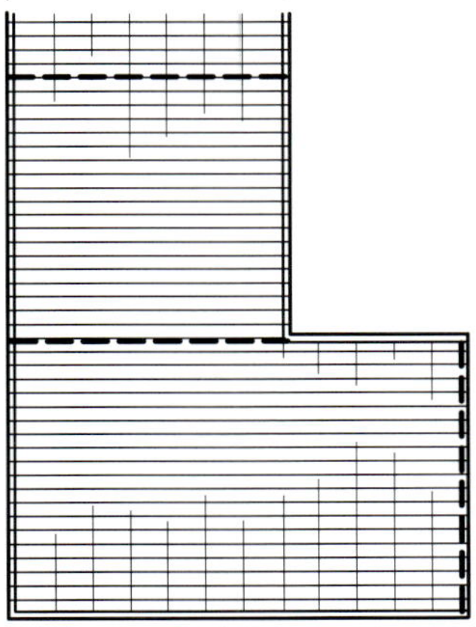

Change of direction at a right angle

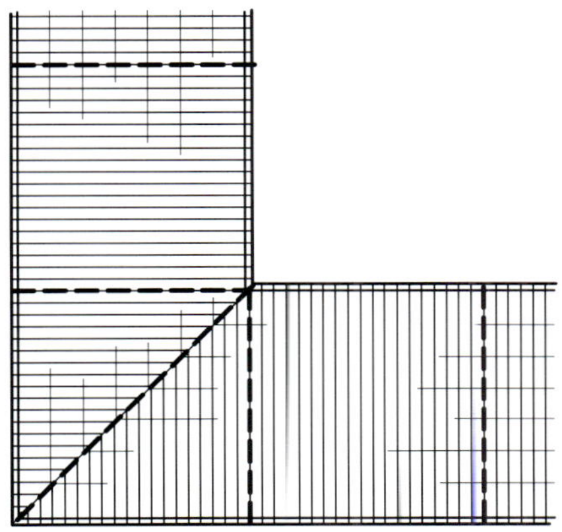

Change of direction on diagonal

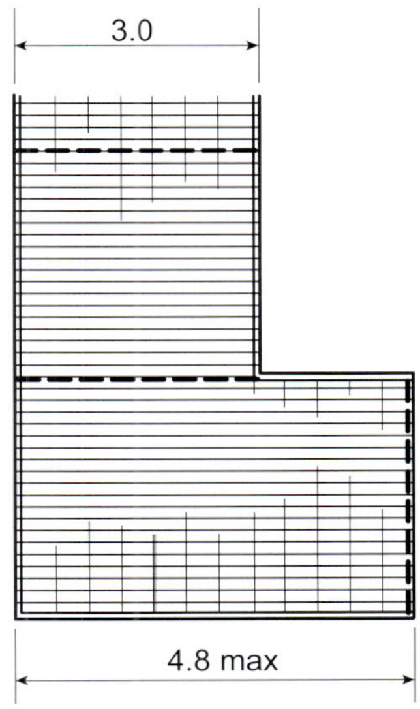

Layout of boards using single lengths only

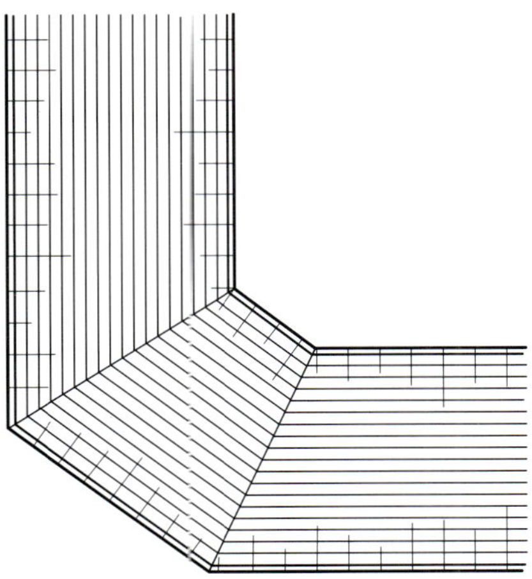

Change of direction using tapered insert

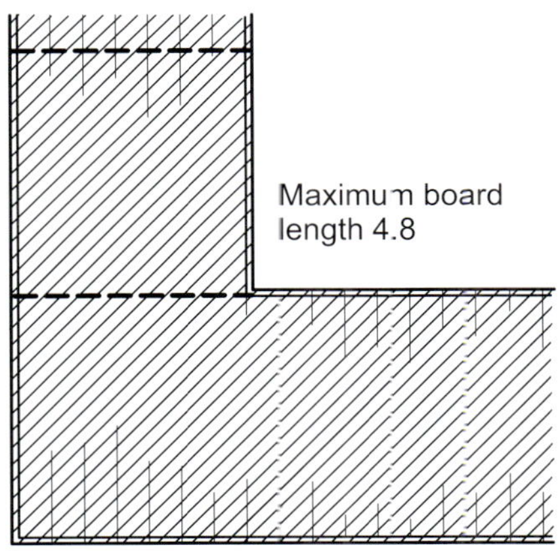

Diagonal board layout

Figure 2.5 Board layouts

Figure 2.6 Boards angled to create decorative shape
Photo: HLD Ltd

If the boards are to be left unfinished, or only finished with a clear moisture resistant coating, the initial colour will bleach out to grey, whether hardwood or softwood is used. Although 'bleaching' may occur at different rates for different species, it is therefore relatively unimportant to choose a wood on the basis of its 'natural' colour.

Clear finishes will not prevent long-term bleaching of the wood and it is necessary to use a pigmented product to provide protection against ultra-violet light which causes the loss of colour. Colours can either simulate the original wood or be a totally different hue. Wood that has been pressure treated with preservative will have more resistance to colour loss even when left unfinished. Preservative treated wood may show a green or brown tinge as evidence of the treatment. If a deck is likely to be exposed to heavy use and abrasion, a tough, durable stain can be used. Regular inspection of the finish is always advisable. If regular maintenance is considered too costly, unfinished wood, either softwood or hardwood, should be considered for both decking and ancillary components. Finishing and maintenance is covered in more detail in Chapter 8.

The overall appearance of the deck will depend considerably on the detailing of the superstucture (discussed in detail in Chapter 6) - balusters, guard-rails, handrails etc - where robust simple detailing often looks more appropriate than the turned or moulded profiles typically used internally. However, there can obviously be a considerable difference in scale between what is suitable for a small domestic patio and a large decked area open to the public. These ancillary components can be treated with the same finishes as the decking or with semi-transparent or opaque wood stains since they are subjected to little abrasion.

To ensure that the long-term appearance of the decking is not spoilt by rust staining, and that the fixings themselves do not deteriorate due to corrosion it is recommended that stainless steel is used for all exposed deck fixings whether screws or nails. Hot dipped galvanised steel can be used for concealed fixings; bolts, clips, plates, brackets and hangers, but there can be a risk of the coating getting damaged during construction and even very limited corrosion can cause staining. Stainless steel is an alternative option for most of these accessories if corrosion is considered an unacceptable risk but they may be difficult to obtain and can add considerably to the cost.

Timber decks will retain their appearance for many years if care has been taken in the choice of materials, the detailing, the construction and the finishing.

2.2 Statutory requirements

2.2.1 Planning control

In general private patios, terraces, pool surrounds, decks etc, as additions to existing properties, do not require planning approval, unless they are nearer to a highway than the existing property or are within 20 m line of sight of a highway, or exceed 3.0 m in height above ground level. Decks attached to listed buildings will require planning approval or if the property is in a conservation area or national park. If there is any doubt, property owners should seek advice from their local planning office.

If a deck is part of the design of a new building, or an extension large enough to require planning approval, the location and size should be shown on the drawings to be submitted to the planning office, but the deck itself would not normally be subject to approval.

If the deck is to be added to an existing building or part of a new development for non-domestic use, such as a pub, clubhouse, hotel, or residential home, it would be advisable to check with the local planning office whether planning approval was necessary. The location and size of the proposed deck should be shown on the planning submission, but would normally only be of interest if it was near a highway or at a raised level. However, if the proposed use might affect the existing amenity of neighbouring properties it would be advisable to discuss this with the planning office at an early stage.

Guidance on planning requirements is available from local planning offices or from the Government's planning portal at www.planningportal.gov.uk.

2.2.2 Building Regulations

For existing properties, a ground level patio, terrace, pool surround etc would not need to be submitted to Building Control for approval under the Building Regulations for England and Wales, or Northern Ireland, or the Technical Standards for Scotland, although if it is directly attached to the property or forms a means of access it is advisable to submit the design. If the deck is to be built at a high level however, it must be designed to ensure that it is stable and capable of taking the loading expected.

If such a raised structure is submitted as part of a new house or other development, it may be necessary to submit structural calculations to show that the design is adequate to take the relevant loading for the particular use.

If the deck is to be part of a development for public use, ie non-domestic, it may be necessary to demonstrate that it will not compromise any designated escape route. If a stair or ramp is part of an escape route it will have to conform with the fire safety requirements of the relevant Building Regulations for the area.

If a change of level of 600 mm or more occurs within the deck area, or between the deck and the surrounding ground, a guard rail should be provided to prevent falling. This should preferably conform with the Building Regulations requirements for guards even if Building Regulation approval is not required.

The requirements for disabled access in the Building Regulations should be met if the deck provides the only access to the primary entrance of the building.

2.2.3 British Standards & codes of practice

To achieve the necessary quality of materials and construction of timber decks it is advisable to refer wherever possible to relevant British Standards or codes of practice. Although none of these specifically covers external decks, many of the materials, and much of the construction, will be similar to those used in building construction generally.

Reference to these standards, at least provides the basis of acceptable quality if any disputes arise over materials or manufacture involved in construction of the deck. A list of relevant documents is included in the References.

3 Performance

The functional requirements of a deck should be considered at an early stage in the design, whether it is for residential or for public use.

It is particularly important that a timber species chosen for decking is appropriate for the type of exposure and use for which the deck is planned. The following criteria are considered the most significant in assessing a species and they are therefore included in the species listings in Appendix I for comparison. These should be taken into account when deciding what species should be used, although cost will also be a deciding factor, and may necessitate some compromise. Also the natural performance of most species can be substantially improved by the use of preservatives, water repellent impregnation or surface coatings as required.

- Durability
- Appearance
- Strength
- Stability
- Resistance to wear and impact
- Slip resistance.

3.1 Durability

Long term durability, as well as the appearance of timber decks, will depend a great deal on the correct choice of materials, good detailing and careful construction.

Because decks are fully exposed to wetting and drying cycles it is of primary importance to choose a wood that is rated "moderately durable" or "durable", from which the sapwood is excluded, or to use wood that has been pressure treated with preservative to enhance its durability. However, durability should primarily depend on good detailing and construction.

Maintaining a low moisture content is critical in ensuring durability as fungal attack will not occur below a moisture content of about 22% in the wood. Even if this moisture content is exceeded for short periods there is still minimal risk of fungal attack. Risk occurs only if the wood remains at a moisture content above this level for a long period of time. In the UK the natural ambient external humidity will not raise the moisture content of wood out of contact with the ground or water above about 20%, and most of the time it will be less than this. Even heavy rainfall will not raise this level, other than on the immediate surface of any exposed timber section. However, if this rainfall becomes standing water trapped against the wood, particularly the end grain, for any length of time, the moisture content of the whole section can be increased.

It is therefore important to avoid creating water traps that can hold water for an extended period, and to ensure that any surface water drains away quickly from horizontal surfaces. Also avoid exposing end-grain to direct wetting, or close contact with wetted surfaces of either porous materials such as masonry, or non-porous materials, such as metals, on which water can lie. This can be achieved either by the use of damp proof membranes between the wetted surface and the wood, or by providing a sufficient air gap to avoid moisture being absorbed by the wood. Such gaps will not only allow any moisture to dry out, but also provide sufficient space for any redecoration that might be necessary.

Generally the most effective method of avoiding a high moisture content in the wood is to ensure that there is sufficient natural ventilation around all component parts of the construction. This is particularly important if a deck is built close to ground level or over water, or may be subject to wetting or splashing from below.

These principles should be applied not only to the deck boarded surface, but also to the substructure and ancillary components, such as handrails, steps, balusters, seats, etc. It is particularly important to cap vertical posts by carrying a handrail over or capping with a flashing. If this is not possible, any uncovered top end should be sloped sufficiently to shed water quickly off the end grain.

Careful selection of suitable timber is a key requirement for durability. Some timber species have naturally decay resistant heartwood and where wood is to be used in applications where decay is a risk, the heartwood of such durable woods can be specified. However, the sapwood of even naturally durable species has little or no decay resistance, and must either be excluded or treated with preservative if long-term durability is desired. Preservative treatment can also provide long-term durability to timber species which are less naturally resistant to decay and for wood exposed to external conditions such as a timber deck, a heavy duty, leach-resistant preservative should be used to provide a long service life against fungal and insect attack.

Timbers are classified into five groups according to the natural durability of the heartwood to decay. (For an explanation of durability ratings see BS EN 350-2.) The groups range from Class 1 "very durable", to Class 5 "not durable". Timbers in Class 2 "durable" and Class 3 "moderately durable" are those most likely to be used for decking without preservative treatment (sapwood excluded). The very durable timbers are likely to be harder, heavier and more costly than is required for most decking projects. Timbers in Class 4 "slightly durable" and Class 5 "not durable" will require preservative treatment. The tables in Appendix I list some species commonly used for, or which have properties appropriate for decking.

Use (or hazard) classes (BS EN 335-1) are used to define the risk of attack by fungi and insects in different environments. Decking components are classified in use (or hazard) class 3 or 4 (occasionally 5), depending upon their role and location in the deck structure, see Table 3.1.

Although most softwood members are treated with preservative, there are examples of the use of moderately durable softwoods, such as imported Douglas fir, or European larch. Both these timbers have the advantage of being suitable for certain components without treatment, providing sapwood is excluded. However, for deck boards on which moisture can lie for some time, the use of untreated moderately durable timbers, rather than preservatively treated timber, needs careful consideration of the likely exposure and usage.

Table 3.1 Use (Hazard) classes (from BS EN 335-1)

Biological Use Class	Definition	Application to solid wood	Examples
1	Under cover, fully protected from the weather and not exposed to wetting	Moisture content less than 20%. Risk of attack by surface moulds and wood-staining or destroying fungi is insignificant. Insect attack possible.	Timber within buildings: internal joinery, intermediate floors, internal roof timbers in pitched roofs
2	Under cover, fully protected from the weather but where high environmental humidity can lead to occasional but not persistent wetting	Moisture content occasionally exceeds 20% and allows attack by wood destroying fungi or disfigurement by surface moulds or staining fungi. Insect attack possible.	Flat roof timbers, ground floor joists
3	Not covered, not in contact with the ground. Continually exposed to the weather or protected from the weather but subject to frequent wetting	Moisture content frequently above 20% thus often liable to attack by wood destroying fungi or disfigurement by surface moulds or staining fungi. Insect attack possible.	External joinery, **deck boards and balustrades**, fence rails
4	In contact with the ground or fresh water and thus permanently exposed to wetting	Moisture content permanently above 20% thus liable to attack by wood destroying fungi and by surface moulds or staining fungi. Insect attack possible above ground or above water.	Fence posts, **deck substructures**, lock gates
5	Permanently exposed to salt water	Moisture content permanently above 20%. Risk of attack by marine borers; above water level risk of attack by wood boring insects.	Marine piling, jetties, sea defences

There are also some examples of Western red cedar used for both framing and decking but, because of the relatively low strength of this timber, larger sections may be necessary to achieve the required structural strength.

Hardwood species which are classed as durable or better can be used without preservative treatment, providing sapwood is excluded.

3.1.1 Preservative treatment

Commercial treatment of wood generally involves combinations of vacuum and high pressure cycles to achieve deeper penetration and higher preservative retention levels than can be attained by other means. The actual treatment used can be varied to match the biological hazard, the wood species being treated and the service-life expectations. The sapwood of certain timber species can be difficult to treat and full sapwood penetration of such species cannot always be achieved, which may be reflected in the service life. Application of wood preservatives by brush or by dipping only provides short term protection to wood and is not recommended.

Timbers that require preservative treatment for decking applications should be treated by a vacuum pressure process using a copper-containing preservative. These modern preservatives have replaced treatments such as CCA (chromated copper arsenate) and CCB (chromated copper boron) which are no longer available in the UK. The copper-based preservatives are water-borne and wood to be treated must be dried to a moisture content no greater than 28%. Above this value it cannot be satisfactorily treated since the wood voids already contain sap or water. As far as possible, all cutting, machining, notching and boring should be carried out prior to treatment.

Wood treated this way generally has a light-green or occasionally brown colour. Treated timber must be re-dried after treatment, to the required moisture level for installation. For most decking timbers, the moisture content should be a maximum of 20% at installation, unless special provision has been made in the design to accommodate the drying movement and possible distortion which may result from installing 'green' or wet timber.

While the pre-treatment ensures adequate protection of the wood, it is still important that any component which is worked on site, either by cutting, notching or boring, should be protected by a compatible brush-applied preservative to prevent direct exposure of untreated wood. Rip sawing, thicknessing and planing should never be carried out after treatment.

Glues and surface coatings are generally not affected by the treatment but it is advisable to consult the preservative manufacture on the use of their particular product. Metal fixings should be made from stainless or hot-dipped galvanised steel.

In addition to enhancing the resistance of the timber to wood-destroying organisms, the preservative treatment also reduces the effect of ultra-violet degradation and slightly improves the surface hardness. Water repellents can be included with the preservative treatment which further improve the weathering characteristics of the treated timber. This causes surface water beading, improves dimensional stability (reducing the risk of surface checking or splitting) and slows the weathering (bleaching) process.

Durable or pressure-treated wood posts can be embedded directly in the ground or they can be supported on metal brackets above ground level. If the posts are to be embedded in the foundation it is essential that a careful choice of species is made or, if the wood is to be treated with preservative, that an appropriate specification is chosen for the required service life. Embedded posts should always be placed with an uncut end in the ground.

Reference has already been made to the need for surface protection on some woods to ensure their long-term performance. Regular inspection and re-treatment will be necessary to maintain effectiveness of such a coating. See Chapter 8.

3.2 Appearance

If left unfinished, all timbers are subject to weathering, which will result in a fading of the initial surface colour, leading to an eventual silvery/grey appearance. The eventual grey colouration will vary depending upon the degree of exposure to sun, the proximity of plants, the liability to sustained wetting of the deck and pollution levels prevalent in the area. This fading results from ultra-violet exposure and woods can react quite quickly, completely losing their initial colour within a period of months. With treated softwoods, the preservative colour does provide some protection and reduces the rate of fade but ultimately the same silver/grey shade will be achieved after several years. Finishes are covered in Chapter 8.

Often, the level of exposure may not be too significant in planning a deck, unless the deck is to be in permanently shaded areas, under heavy foliage, or exposed to exceptional wetting, eg pool surrounds. If a deck is to be left exposed to weather naturally the rate of bleaching may be inconsistent between areas fully exposed to sun and those generally in shadow, or shaded by projections.

Applying a finish which reduces the moisture permeability of the wood will reduce the risk of the boards being deeply stained by bird droppings, fouling from pets or other liquids. For public decks, where there may be a higher risk of spillage, either accidentally or as a result of vandalism, it may be preferable to choose a denser wood as this can be more easily cleaned by detergents or chemicals where necessary. Wood is generally more resistant to chemicals and pollution than other building materials.

Close proximity to fresh water, with a higher risk of wetting and possibly a higher humidity if the deck is built close to the surface of the water, should be taken into account when choosing the species and any finishes. Wetting from seawater in the form of spray or splashing is actually beneficial in that the salt tends to inhibit the growth of algae. While it is always advisable to use stainless steel or hot galvanised fixings and fittings for decks, this is particularly important in a maritime environment, where there is a high risk of corrosion and consequent risk of staining to the wood.

For structural timbers, the strength grade chosen (C16 etc) will limit the size and spacing of knots, splits etc. For deck boards the standard for timber in joinery, BS EN 942, can be used to limit knot sizes and spacing for appearance reasons.

3.3 Strength

Any deck will need to be sufficiently robust to sustain the likely imposed loads. In a domestic deck the loading will probably only consist of a limited number of people, and fairly light point loads, such as from garden furniture. However spaced, deck boards do not spread loads in the same way as tongued and grooved floorboards. While a similar joist size and spacing and board thickness may be theoretically adequate to support the expected loads, the deck may prove too flexible, which can prove disconcerting in use.

If the deck has been designed for normal domestic use limited to pedestrian loads, it should be remembered that adding excessively heavy objects might exceed the loads for which the structure was designed. For example, large plant pots and deck furniture can be surprisingly heavy and should be positioned with care. Pools, spas and hot tubs should never be supported on a deck structure unless it has been specifically designed for this purpose. If in doubt about any of these aspects it is recommended that the deck supplier and constructor be consulted.

Where decks are designed for public use, such as pub or clubhouse terraces, shopping malls, or locations where they may not be exclusively for pedestrian use, it may be necessary for the deck to take much higher loads, including rolling loads, such as cycles, golf buggies, wheelchairs or boat trolleys. On larger public areas it may also be necessary to allow for the weight of cleaning and maintenance vehicles.

In these cases it will be necessary to design the deck and substructure to accommodate this loading, particularly from wheeled vehicles which may impose short-term dynamic loads. For external use, somewhat higher short -term deflections are normally considered acceptable than would be permitted for internal floors, but the deck should not be too 'springy'. The design of these decks should be checked by a specialist engineer.

For decks for public use with higher loadings, as well as increased impact and wear, hardwood boards might be a preferred alternative to softwood because of their greater strength and higher density.

For all decks it is generally preferable to limit the spacing of the supports to a maximum of 600 mm, rather than to increase the deck board thickness in order to achieve larger spans. Limiting the spacing of supports is as much to do with keeping the boards straight and flat as with the spanning capacity of the deck boards. Hardwood in particular is generally more expensive than softwood, and increasing the thickness can disproportionately increase the cost compared to adding extra joists at closer spacing. However, there are some situations where longer spans are necessary and in this case thicker boards are readily available in both softwood and hardwood.

3.4 Stability

All wooden components will have some tendency to develop surface splits, bow, cup, twist, or distort sideways as the moisture content varies depending on where they are cut from the log, regardless of the section size, and even if the wood is rated as one of "small" or "medium" movement. More closely spaced joists, while limiting the span of the boards, will also reduce the spacing between the fixings, and these fixings will help restrain movement of the board. Any stresses that may develop will also be less severe in thinner boards and more easily restrained by the fixings.

In some cases, deck boards are used 'undried' or 'green' in order to save the cost of kiln or air drying. In this case it is essential to fix frequently along the length of the board and also to allow for any shrinkage across the board width as it dries. This situation is typical of temperate hardwoods, such as European oak, but wood cut from large imported tropical logs can also be initially at quite a high moisture content. To reduce the effects of shrinkage across the board as much as possible it is advisable to use relatively narrow boards, preferably no wider than 150 mm.

To limit the movement of boards in use, they should preferably be installed at a moisture content of about 20% maximum. There will always be some variation in the moisture content during the course of the year as a result of changes in the relative humidity. To avoid heavy rainfall raising the moisture content of the surface of the boards, water should be drained away quickly between wettings. Softwood boards will generally absorb moisture more readily than hardwood, but there are various treatments available to make them more moisture resistant and consequently more dimensionally stable.

Small surface cracks or 'checks' often occur in deck boards due to drying out when exposed to hot sun. These will usually close up when the surface is wetted, but it is advisable to choose woods that are rated as having "small" or "medium" movement, as there will be less risk of checks or splits developing as the surface moisture content is reduced.

3.5 Resistance to wear and impact

Deck boards are subject to gradual wear and erosion over a period of time depending on their use. Care should be taken both in the initial choice of a timber to stand up to the anticipated end use and in the method of fixing if some boards may need to be replaced in the long term.

As a general rule, timbers of lower density will be subject to greater wear than those of higher density. In practice however, for all private, domestic and many commercial deck schemes, good quality softwoods can offer a satisfactory resistance to wear as proven over many years in such applications.

Applications where dense hardwoods do offer some advantages include areas where studded or spiked sports shoes are worn or where high impact or high point loads are anticipated. Because such severe usage would quickly damage any surface finish in these locations, it is preferable to leave the hardwood unfinished.

For more typical conditions of use a surface finish can provide some wear resistance to the surface of the wood, but heavy local wear can lead to an uneven appearance which may require more frequent re-coating of these areas, or of the deck as a whole.

3.6 Slip resistance

It is a common perception that timber decks are slippery and potentially hazardous when wet. Whilst there is certainly a potential for this, the risks are often overstated and the problem can be eliminated, or minimised, with careful detailing, choice of appropriate materials and routine care and maintenance. The main areas of risk are ramps, steps and landings at the head of steps, or immediately adjacent to doorways.

The primary and most common cause of slipperiness on timber decks is standing water, which can drastically reduce underfoot grip and can also cause the timber to become saturated and inherently more slippery than dry wood. Practical steps can be taken to ensure that standing water does not occur. These include ensuring that decks are installed level without any low points or, preferably, installed with a positive drainage fall. The extent of fall need only be slight and this would probably not be detectable to the user, but it is important that falls run away from a building to the outer edge of the deck or to a drainage channel provided.

A secondary cause of potential slipperiness is the occurrence of algae, moulds and slime growths which can occur on timber which is waterlogged or which remains wet or damp for extended periods. Areas of decking in constant shade or under tree or plant cover are particularly susceptible and where such growths occur they can lead to extreme slipperiness, particularly in wet conditions. The problems can be particularly pronounced in the autumn period.

The incidence of algae, mould and slime growth can be minimised by careful positioning of the deck to avoid areas in constant or prolonged shade. Similarly, overhanging trees and adjacent plants and vegetation should be controlled to prevent constant damp conditions. Occasional surface applications with anti-fungal solutions can limit organic growth on decks subject to these conditions.

The final cause of slipperiness is a seasonal problem of surface frost and ice on the deck boards. In this respect timber is no different from any other walkway material and shedding surface water through drainage is the prime means of reducing the problem. Unfinished timber will, nevertheless, absorb some surface water which could result in slipperiness in icy conditions. The use of water-repellent treatments or finishes will reduce moisture absorption into the timber but may risk surface icing.

If surface frost or ice does form, standard rock salt can be used to alleviate the problem. However it should be pointed out that this could transfer salt indoors on shoe soles and in addition some surface staining of the deck boards may occur, although this can usually be removed later by cleaning in the spring. The use of rock salt could also have a detrimental effect on previously applied coatings.

The use of sawn, textured or grooved boards will improve slip resistance over plain boards but consideration should always be given to the 'comfort' of the deck finish, particularly in a domestic situation where users may be barefooted.

Sometimes very dense hardwoods are specified for strength or wear resistance in heavily trafficked areas, but these woods can become highly polished with use, resulting in increased slipperiness. Less dense hardwoods will roughen with age, possibly improving their slip resistance.

When laying decking it is important to allow an adequate drainage gap between each board rather than butting the boards together, see Section 5.2.

Grooved boards are popular and widely available with a variety of differing profiles and patterns, see Section 5.2. The main purpose of the grooves is to provide drainage channels to remove surface water and therefore it is important that the spacing, depth and width of the grooves are adequate and that they do not become blocked with debris. The boards should be laid with the grooves in the direction of fall.

Where there is considered to be a higher risk of slipping, such as ramps, steps or landings on decks used by the public, specialist decks coatings are available in the form of clear sealers or varnishes with sand, pumice or mineral grit to improve traction. To maintain their performance these coatings need to be regularly re-applied in accordance with the manufacturers recommendations. Alternatively boards are manufactured with non-slip inserts installed or bonded into grooves, see Figure 3.1.

Figure 3.1 A variety of non-slip inserts in grooved boards
Photo: TDA/DesignADeck

Routine cleaning, care and maintenance of the deck structure will remove dirt and surface growths and more information is given in Chapter 8.

4 Substructure

4.1 Preparing the site

If a timber deck is to be built well above ground level, site preparation will only consist of digging the foundations for supporting posts, piers or walls.

However if the deck is at a low level, particularly if it is within 600 mm of the existing ground level, it is important to ensure that any timber substructure is clear of the ground and not in direct contact with vegetation. Besides wetting the wood, plant growth under the deck will tend to restrict ventilation and drainage. Open jointed boards will allow sufficient rainwater and daylight to filter through to encourage the growth of weeds and therefore positive action should be taken to prevent this.

It is normally adequate to remove all existing growth and at least some of the top soil. The surface should then be lightly compacted and laid to at least 1:40 fall towards the outer edge of the deck to ensure drainage away from any structure to which the deck is attached.

A weed killer can be added to sterilise the soil, but usually laying a weed suppressing membrane over the whole area under the deck will be sufficient to prevent growth. This sheeting should be held in place by 50 mm of gravel, which will drain readily and discourage growth above it.

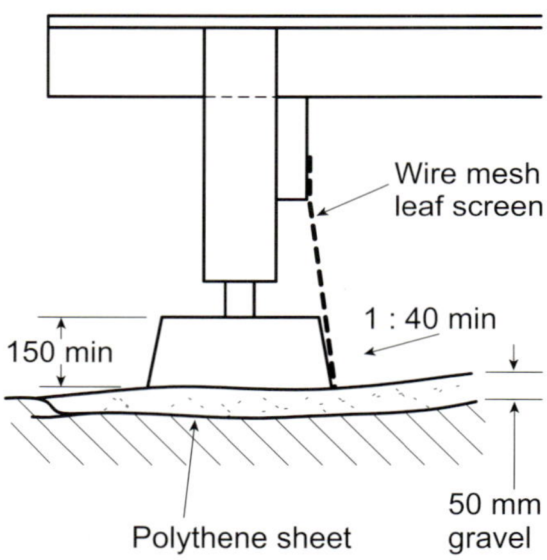

Figure 4.1 Site preparation

Despite inhibiting the growth of vegetation, it is probable that dried leaves and other rubbish which could constitute a fire hazard may accumulate under a low deck over a period of time due to wind. This can be reduced by installing a vertical metal mesh or plastic net with openings not larger than 12 mm x 12 mm which should be located sufficiently far under the deck so that it is not visible. However a mesh might make it more difficult to clear any rubbish that penetrates through the joints in the decking from the surface unless it can be taken out when cleaning is necessary. If leaves and rubbish are allowed to build up against this barrier it could itself restrict ventilation and drainage from under the deck. See Figure 4.1.

4.2 Substructure support

The timber substructure for decks is generally supported on posts, columns or piers. These supports are usually timber posts or poles set in, or on, concrete footings, but can also be steel, precast concrete or brick piers.

Figure 4.2 Top of post sloped to shed water

Timber is the most commonly used material for the supports, either as posts or round poles. Rectangular posts can be of solid wood or laminated sections. All posts must be a durable hardwood or softwood preservative pressure treated to Use Class 4 standard, see Table 3.1. In this case, it is particularly important that if the posts are cut to length, the cut ends are protected with a compatible end grain preservative.

The spacing of these vertical supports will depend on the spanning properties of the beams. The size and design of these beams will be determined by the overall loading, and stability of the structure. Stability becomes more important the higher the deck is above ground level, and it may be necessary to add cross bracing in some cases.

Alternatively the deck substructure can be directly supported at one end on a wall plate or hangers fixed to a masonry wall, rather than on beams. Brick or stone walls can provide good support for a limited height but they may not be sufficiently stable for a high level deck and will require more substantial foundations than required for posts.

Figure 4.3 Post on standard steel support

Steel "H" sections, or hollow tube sections, can be used for columns and have the advantage of providing a visually light structure when supporting high level decks. Exposure to the weather can cause corrosion and the sections should preferably be galvanised to reduce the need for regular repainting. Steel columns can be either embedded in concrete or bolted to the top of concrete foundations.

Precast concrete piles are occasionally used as columns. For low level decks precast concrete pads, or poured concrete in precast concrete rings on concrete bases provide a simple support system. Specialist foundation former tubes may also be used. Timber posts can be bolted to steel 'shoes' to hold them clear of concrete foundation pads and there are a number of types available, see Figures 4.3 and 4.5.

If timber posts are to be embedded in poured concrete foundations it is preferable to either use concrete without a sand aggregate throughout ('no fines'), or to form a hole in the standard concrete base larger than the post so that after it is inserted a dry mix or gravel packing can be added to hold the post firmly in position. For either method the bottom of the foundation should be left open. This is to ensure that there is good drainage around the wood, and that the column does not stand permanently in water, see Figure 4.4. If the posts are to stand in water, or there is a very high water table, different foundations details will be required.

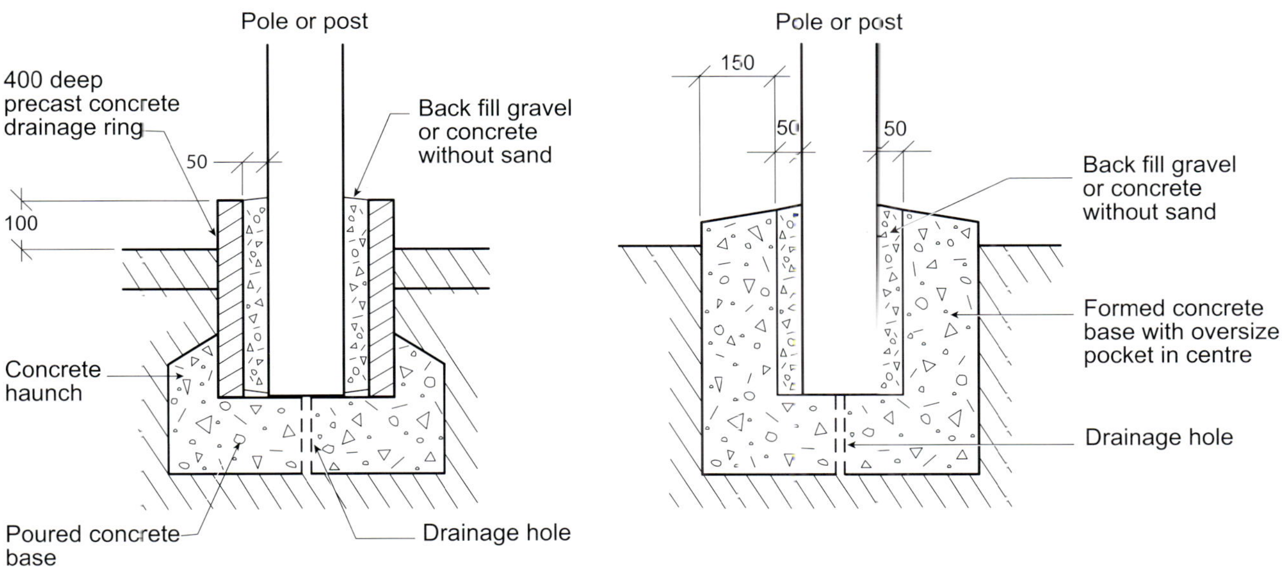

Figure 4.4 Concrete foundations. Two methods of embedding posts or poles

Figure 4.5 Two methods of mounting posts on steel brackets on concrete bases

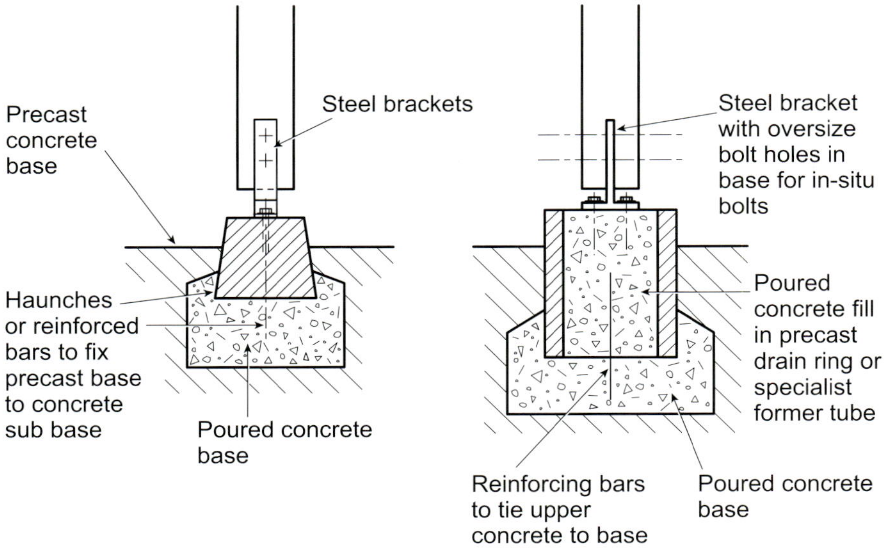

4.3 Deck substructure

Timber decks are usually fixed to a timber substructure consisting of joists spanning between beams which are directly supported by the posts, piers or cross walls. The main advantage of using timber is that all connections can be made by simple carpentry joints or with standard metal accessories fixed with nails, screws or bolts. Where a guard rail is required, the posts or poles can often be extended to provide support for the rail.

If the beams are to be of timber, there are some advantages in using two members rather than one as the reduced section thickness will minimise the effects of any moisture induced movement in the timber. For economy, single beams can then be used at the edge of the deck, rather than double as these only carry half the deck load, although double beams are more common. The beams can simply be bolted to the side of any posts, either at the same height or at different heights to accommodate any changes of level of the deck. This arrangement also allows for easier extension of the deck in the future, if this is found necessary. Typical details are shown in Figures 4.6-4.8.

Joists supporting decking are usually carried over the top of the beams. This has the advantage that the joists for wider decks can be designed to span across a number of beams, possibly reducing their required section size. It is normally adequate to nail the joists to the beams, but special steel clips are available and provide more stability when deep narrow joists are required.

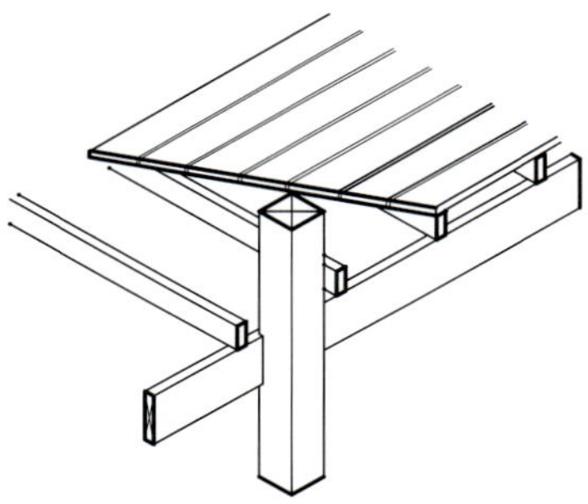

Perimeter parallel to beam, posts projecting

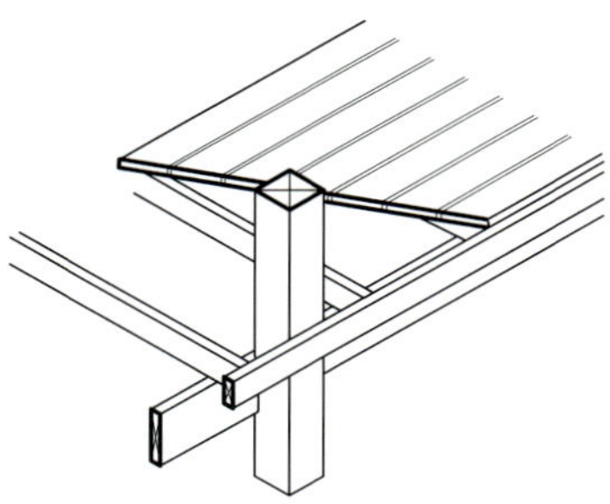

Perimeter parallel to beam, posts within deck

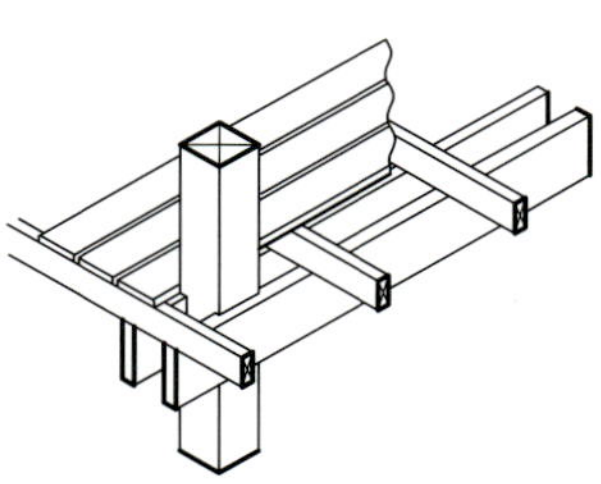

Perimeter parallel to joists

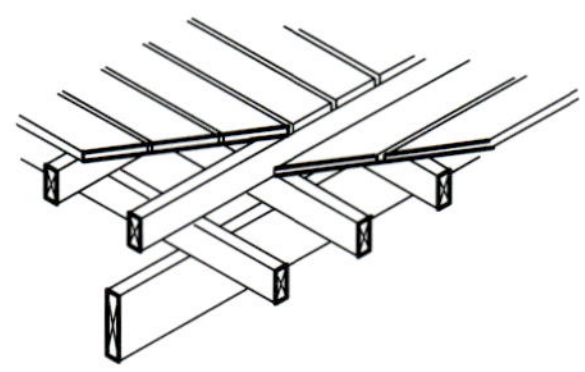

Small change in level

Figure 4.6 Typical substructure details

Figure 4.7 Double beam supporting joists Photo: Gibson Design & Build

Another advantage of carrying the joists over the top of the beams or wall is that, if required, the joists can be cantilevered out beyond the supports. As a rule of thumb, joists can be cantilevered out a distance equivalent to about a quarter of their clear span. See Figure 4.9. Projecting the deck out over the supports can visually 'play down' the support structure but also has the practical advantage of keeping the foundations well below the deck. This can be particularly advantageous on sloping sites, as moving the supports up the slope will reduce their height.

Figure 4.8 Double beams supporting joists

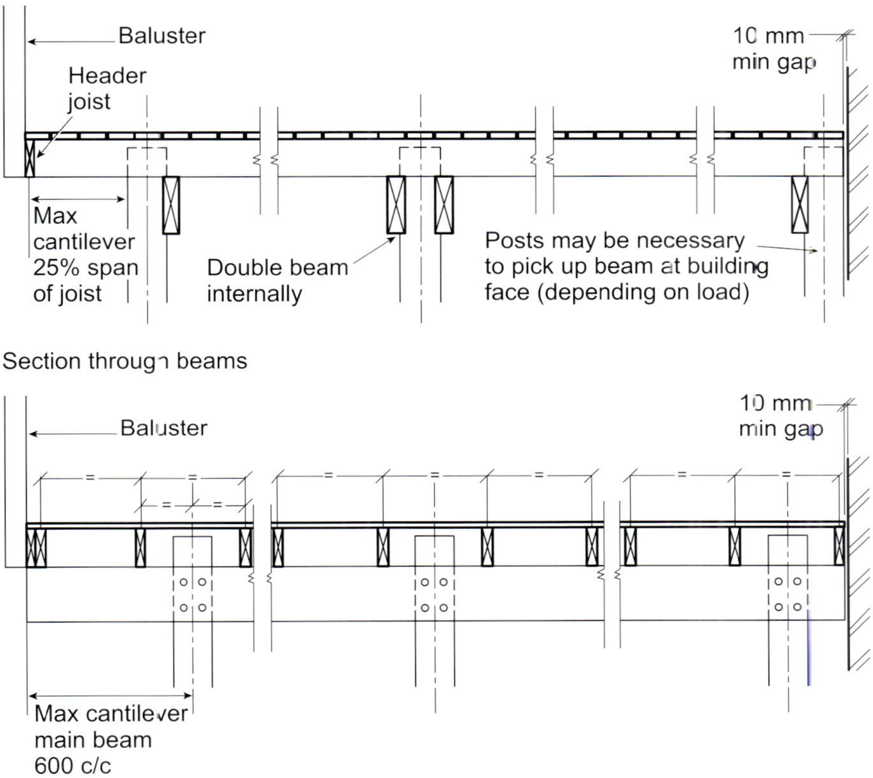

Figure 4.9 Projecting joists over beams

If it is necessary to limit the overall depth of the deck construction the supporting joists can be set between the beams so that their tops are at the same level. In this case it is not practical to project the joists beyond the edge supports. The joists can be supported on galvanised or stainless steel joist hangers fixed to the sides of the beams, see Figures 4.10 and 4.11b. The type of joist hanger that clips over the top of the beam is best avoided because the top flange can conflict with deck fixings which lie on the joist lines. Alternatively, the joists can be supported on ledgers, which are continuous battens screwed to the beams, see Figure 4.11a. These will require structural calculation. The tops of ledgers should preferably be chamfered to shed water and the ends of the joists should be cut to the same pitch to ensure a good bearing.

Figure 4.10 Standard pressed steel hanger

Where joists are set level with the top of double beams, blockings must be added between the beams at joist centres to ensure load-sharing. The blocking should be of the same depth as the joists, and if level, will provide support and fixing for boards that occur between the beams, see Figure 4.11.

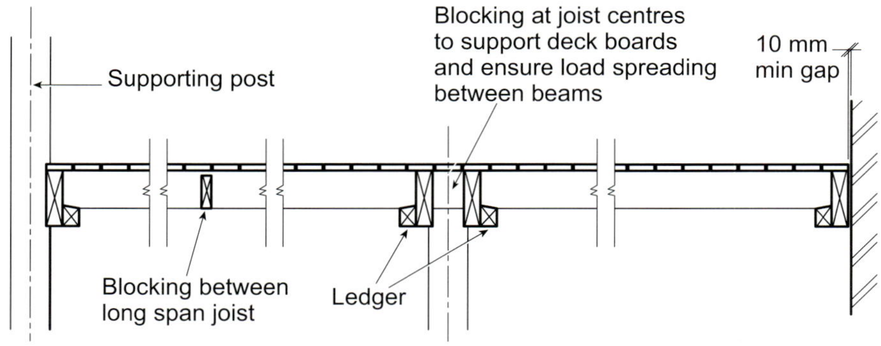

Section through beams

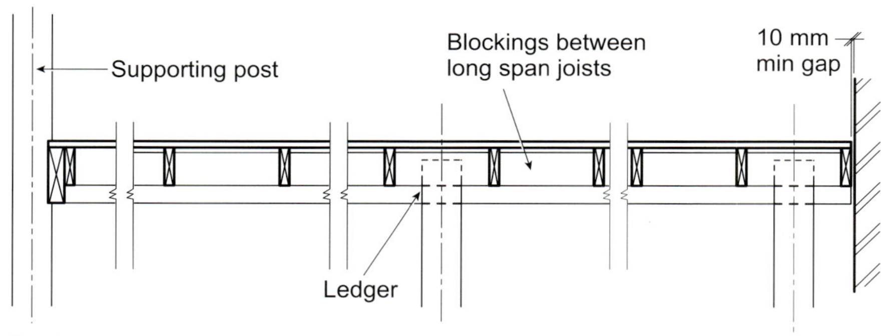

Section through joists

Figure 4.11a Joists on ledgers between beams

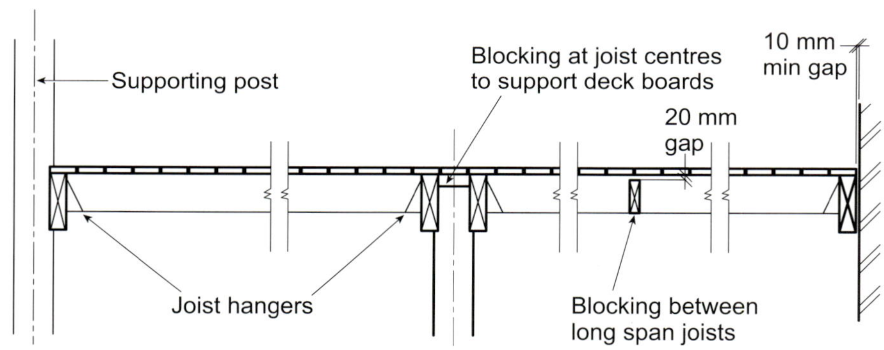

Section through beams

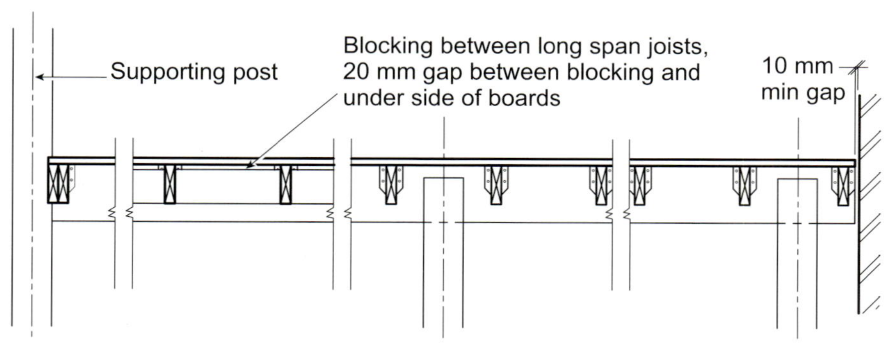

Section through joists

Figure 4.11b Joists on joist hangers between beams

Joists above 2.5 m span will require blocking at mid-span. This blocking should be kept at least 20 mm below the tops of the joists to allow ventilation and drainage. See Figure 4.12.

In the design of any substructure it is preferable to provide more frequent supporting joists rather than to increase the thickness of the decking boards to achieve larger spans. Closing down the spacing of joists means that there are more fixings along the length of the board which will help to limit moisture-induced movement. The spacing of the supports should normally be limited to a to a maximum of 600 mm (see Appendix II Span tables).

Where a deck is attached to the side of a building, a wall plate must be bolted to the masonry to support the joists directly, or to carry joist hangers. It is important that this plate is held at least 10 mm away from the face of the wall so that any water running down the brickwork can pass behind the plate, rather than accumulate on the top edge. This gap can be provided by spacer packs or washers under the bolts, see Figures 4.13 and 4.14. Alternatively a continuous metal flashing can be keyed into a mortar joint and dressed down over the top of the plate to keep water off the top surface.

Where joists run parallel to the face of a building they should also be kept clear of the building face by at least 10 mm. The same gap should be provided between the ends, or side, of deck boards and any face of the building, see Figure 4.14. It is also important to ensure that any beams, joists or ledgers attached to the walls do not block existing air bricks.

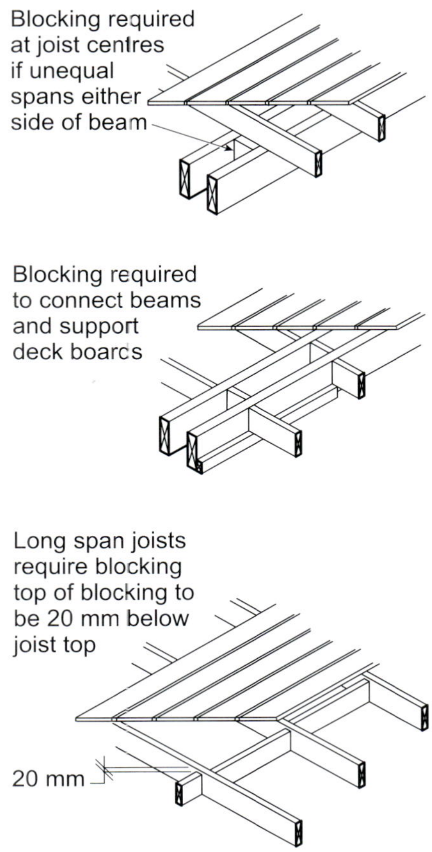

Figure 4.12 Blocking details

4.3.1 Selection and specification of timber

The majority of timber decks are supported on softwood beams and joists, which have been treated with preservative as required. Whichever softwood species is chosen, it should be structurally graded as C16 or better. Higher strength timbers such as C18 or C24 can sometimes be useful in larger decks, in that using them can reduce the section sizes (see Appendix II Span tables). This may be an advantage if there is limited space for the substructure.

Hardwood members are occasionally specified in association with hardwood decks, usually European oak, but occasionally of tropical hardwoods. If European oak or sweet chestnut are to be used for any part of the substructure, they should be graded to TH1 under BS 5756 (see Span tables). If a tropical hardwood is to be used, it should be to at least HS grade.

Figure 4.13 Beam attached to wall: note gap with spacer packs

4.3.2 Substructure fixings

While stainless steel is recommended for fixing boards to the substructure, this may prove too expensive to use generally in the construction of the substructure itself. Generally coach screws, nuts and bolts, washers and screws should be at least of hot-dip galvanised steel. This also applies to column supports, flitch plates, angle cleats etc. While joist hangers are available as galvanised mild steel, stainless steel joist hangers are commercially available and are relatively inexpensive. Because, unlike internal floors, the hangers are permanently exposed to the weather, the additional cost of using austenitic stainless steel may be considered worthwhile. If the hangers are to be of stainless steel, the fixing nails should also be stainless, but stainless steel fixings must not be used in conjunction with galvanised sections because of the risk of electrolytic action between the metals.

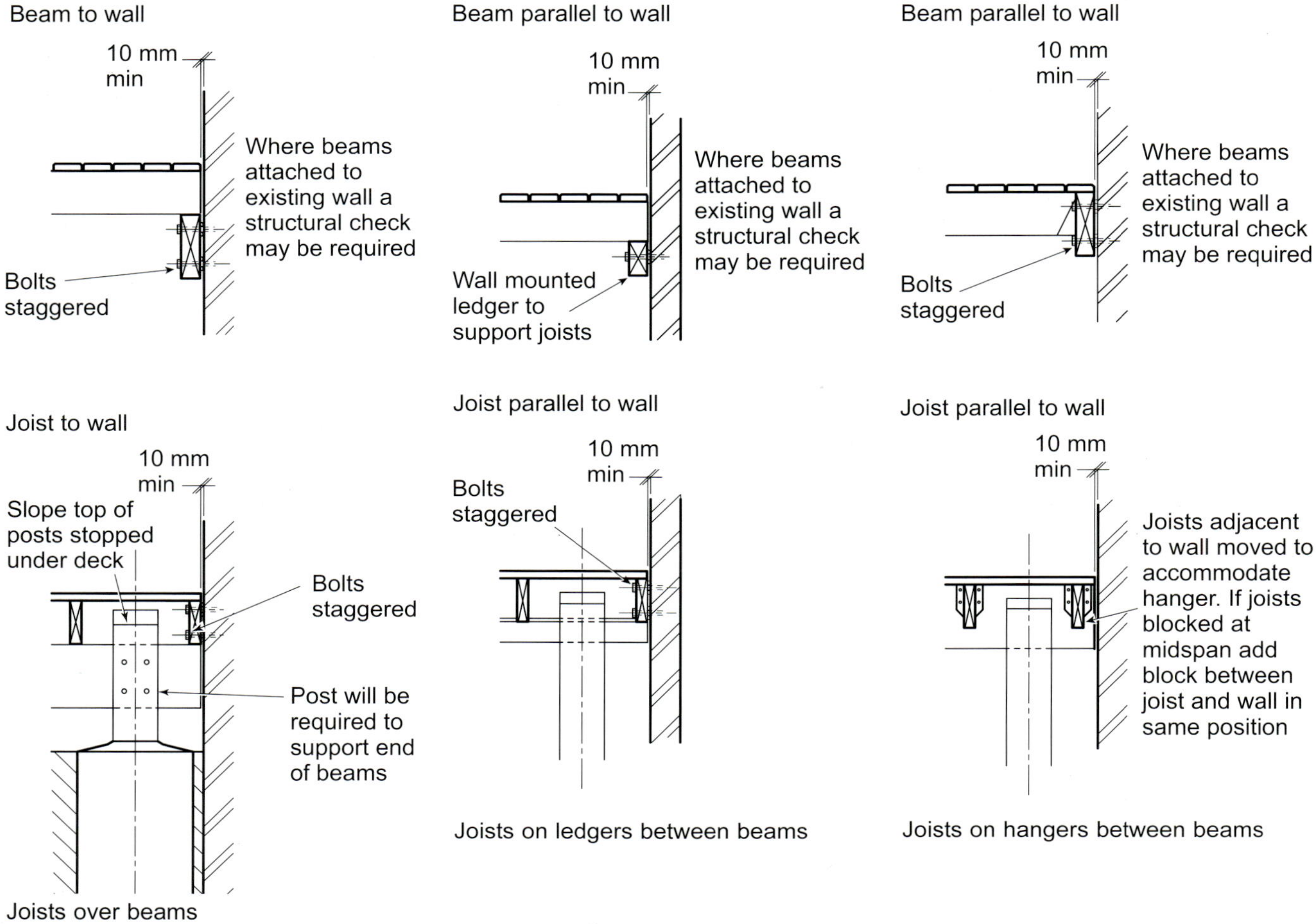

Figure 4.14 Attaching to existing buildings

For additional protection, and for appearance, most steel members in a deck are painted, but it is important to use a high quality metal paint because of the level of exposure expected. It may also be difficult or impossible to repaint metal parts if they are inaccessible after the deck is completed.

Care should be taken if any self-drilling steel fixings are used into steel members, because the swarf can corrode and stain timber members, particularly if wood with a high tannin content, such as Western red cedar or oak, is used.

Figure 4.15 Decking over existing roof
Photo: HLD Ltd

4.4 Adding decks over existing structures

In some cases where it is intended to add decks over an existing paved area, or flat-roof, it is necessary to keep the change of level from an existing floor to the deck to the minimum. In this case, a separate framed substructure can be omitted if support can be obtained directly from the existing base structure and this is capable of taking the additional loading. The surface of this base should be sloped to drain any water that runs through the gaps in the decking away from beneath the deck. If any deck bearers sit directly on the base they are likely to restrict drainage and may also be standing in water for prolonged periods. The bearers should be notched to allow drainage or preferably held off the surface of the base by plastic or durable hardwood packs at least 12 mm thick. There are a number of commercially available adjustable plastic supports designed for paving slabs that can be adapted for this purpose. The bearers in this case should be designed to span between these supports which are usually placed at about 600 mm centres. These supports can also be used to take up any unevenness in the base level or to achieve a flat deck over a sloping base, see Figure 4.16a.

For wider decks it is preferable to used tapered bearers to avoid the need for a height of packing that might cause the bearers to become unstable. The bearers should preferably never be less than 50 mm thick at their thinnest end, so that there is always sufficient gap between the underside of the decking and the base not to be blocked by rubbish, see Figure 4.16b. This will also allow sufficient ventilation to the supports and underside of the deck. Only durable hardwoods or preservative treated softwood should be used for the bearers. Some thought should be given as to how any rubbish leaves or dirt that may accumulate under the deck can be removed periodically. One possible method is to fix the boards to battens to form a panel similar to a 'duckboard' which then can be removed for cleaning underneath.

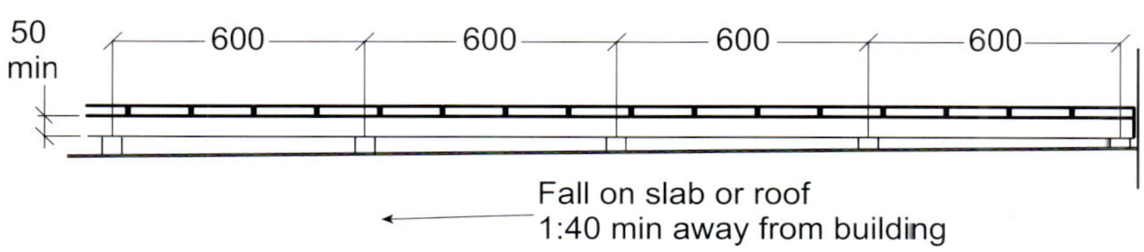

a 50 x 50 treated battens on plastic or durable hardwood shims, varied in size to suit, low spots etc

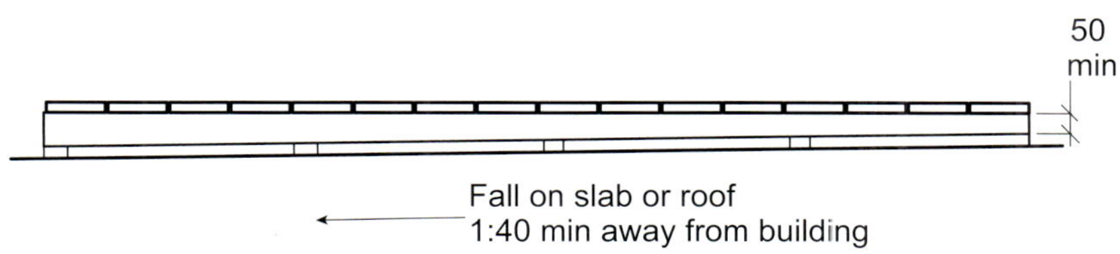

b Tapered treated battens (min thickness 50 mm) on plastic or durable hardwood packs at 600 mm centres

Figure 4.16 Decks over existing slabs or flat roofs

As an alternative to laying boards to give the appearance of a monolithic deck, there are now various prefabricated decking panels available, either square or rectangular, that can be supported directly on a loadbearing base. These consist of treated softwood decking boards fixed either square or diagonally to a supporting frame of treated softwood. These panels can be laid either with the boards in line or in a chequerboard pattern and they are supported on levelling packs or bearers, see Figure 6.17. The supports should be sufficiently high to allow drainage of any surface water from underneath. The panels should have sufficient weight to resist wind uplift or be mechanically fixed to the substructure or roof, if it is acceptable to penetrate the roof membrane.

An advantage of this method is that the layout of these panels can be very flexible and it is simple to form small changes in level by overlapping the panels on top of each other.

Figure 4.17 Prefabricated decking panels
Photo: HLD Ltd

5 Deck boards

5.1 Board dimensions

It is advisable to limit the width of deck board for the following reasons; to reduce the effects of moisture movement, to ensure quick drainage, and to improve traction by increasing the number of gaps between boards. The maximum nominal width of board should not exceed 150 mm. Common nominal widths are 125, 100 or 75 mm, see Figure 5.1. If the board is 75 mm wide or less, only a single fixing is required in the centre of the board. If the wood for the deck is to be used undried a single fixing has the advantage that any shrinkage will not result in stresses developing between fixings which may cause splitting.

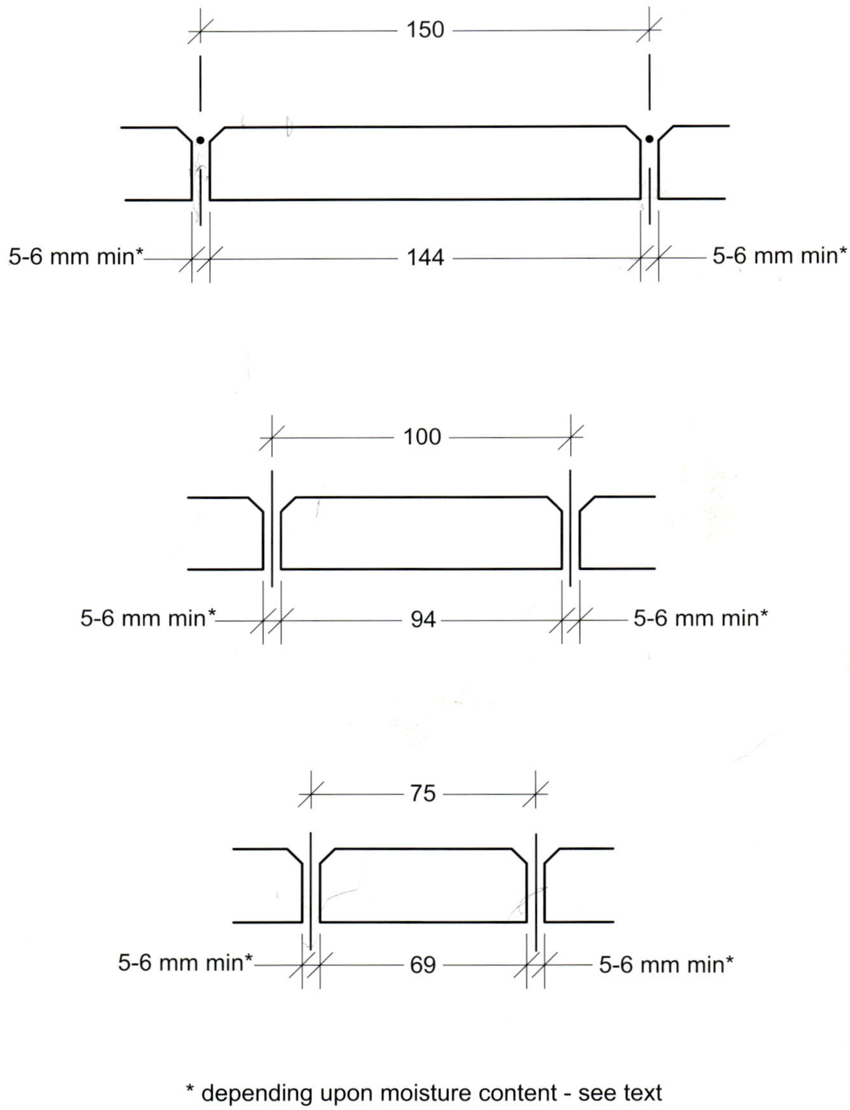

Figure 5.1 Typical board dimensions and profile without drainage grooves

5.2 Board profiles and spacing

Whether plain, ribbed, or grooved boards are used the top edges of all boards should be radiused or chamfered. Any surface water will run off a chamfer or radius more quickly than off a sharp edge and will also prevent any surface finish 'thinning' at the transition between vertical and horizontal faces. A chamfered or rounded edge is less likely to split off in use due to impact and if a board does distort, a radiused or chamfered edge will tend to remain relatively flush with the adjacent board and not become a tripping hazard.

Two or three sawcuts per board width and approximately 3 mm deep on the underside of the boards are sometimes added to help prevent any tendency for the board to cup or bow by reducing tension stresses on the underside of the board.

Efficient drainage of decking will depend on the width of the gap between the boards. If this is too narrow water will be held there by capillary attraction, or the gap will be blocked by dirt. If water is held for either reason there is a risk that the wood will swell and reduce the gap still further.

The gap should not be less than 6 mm to avoid water droplets being held, and should preferably be at least 8 mm wide to avoid dirt accumulating and to ensure that water can flow away rapidly. If the boards are installed at a moisture content higher than they are likely to reach in use, the gap can be 5 mm on the assumption that it will increase as the moisture content reduces. A maximum gap of 10 mm is suggested to avoid problems with narrow shoe heels, bicycle or pram wheels, roller blades etc. Where there is a high risk of dirt accumulating in the joint, widening the gap below the surface of the boards by cutting the side of either one, or both boards at an angle will reduce the risk. The consequent V-shaped space will reduce the amount of dirt held without the need to increase the width of the gap at the top surface. See Figure 5.2.

Grooves in the surface of boards are intended to carry away surface water, improving the slip-resistance in the process, and must be adequate for this purpose without being easily blocked by dirt, see Figures 5.2 and 5.3. If the groove is too narrow, or too shallow, it will block easily, and also be difficult to clean. If the groove tapers in section to form a 'wedge profile' it will not hold dirt and can be more easily cleaned. Grooves are normally spaced about 10 mm apart and arranged in the width of the board so that there is a flat section of board each side. The edges of this flat surface should be chamfered or radiused into the side of the board.

If non-slip inserts are to be fitted to the grooves because there is a higher than normal risk of slipping eg ramps or stair treads on decks for public access, the groove profiles may need to be changed to hold the strips securely and provide a better mechanical key.

Grooved boards are most effective when the grooves run across the short dimension of the deck as they will drain more effectively. Aligning the boards, whether plain or grooved, across the primary direction of travel will also improve grip on both flat and sloping surfaces.

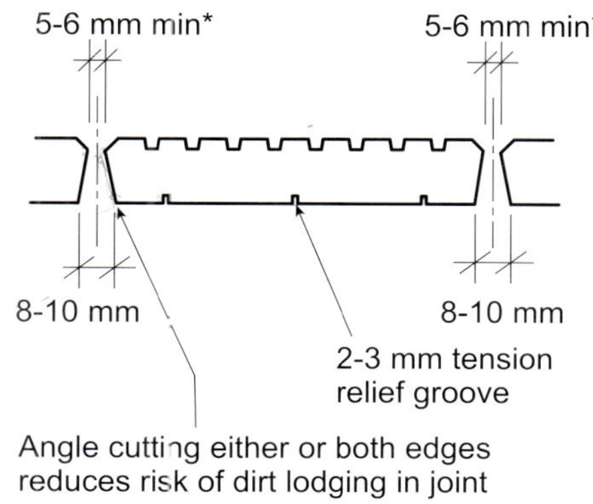

* depending upon moisture content - see text

Figure 5.2 Grooved board profile with angled edges

Figure 5.3 Grooved profile deckboards

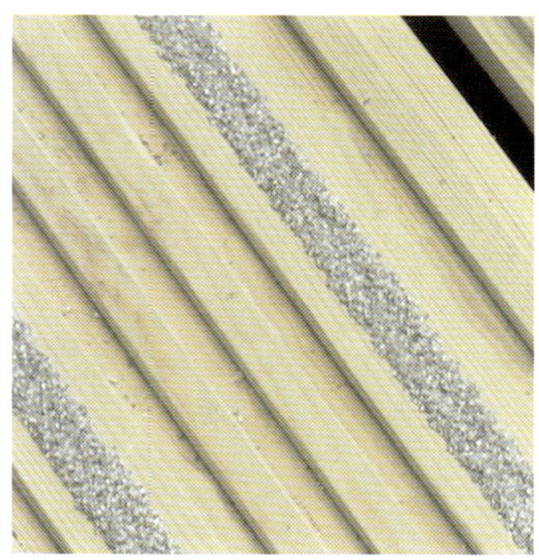

Figure 5.4 Anti-slip inserts in deck boards
Photo: HLD Ltd

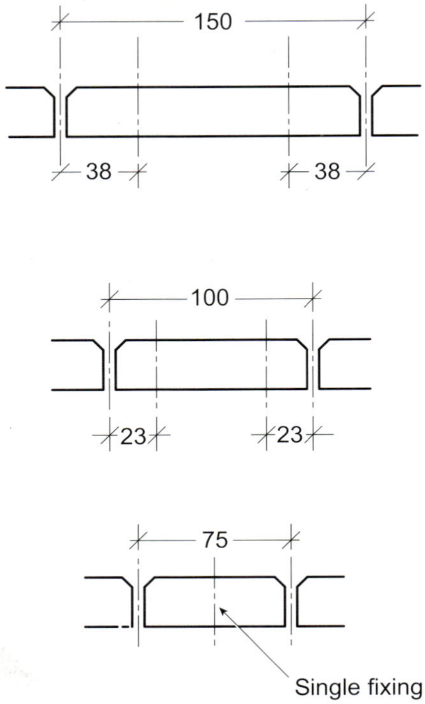

Figure 5.5 Location of fixings

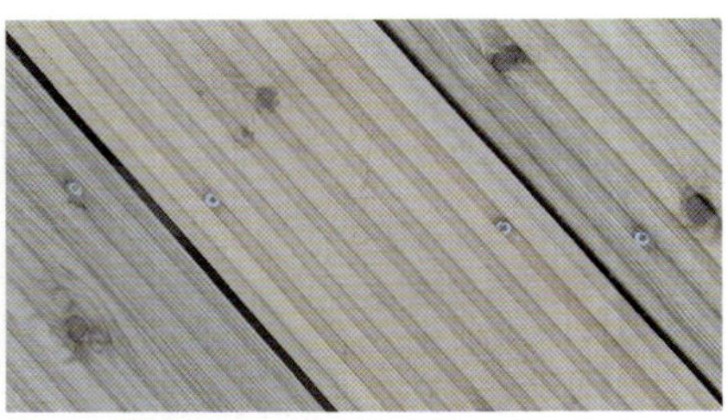

Figure 5.6 Fixings inserted within grooves

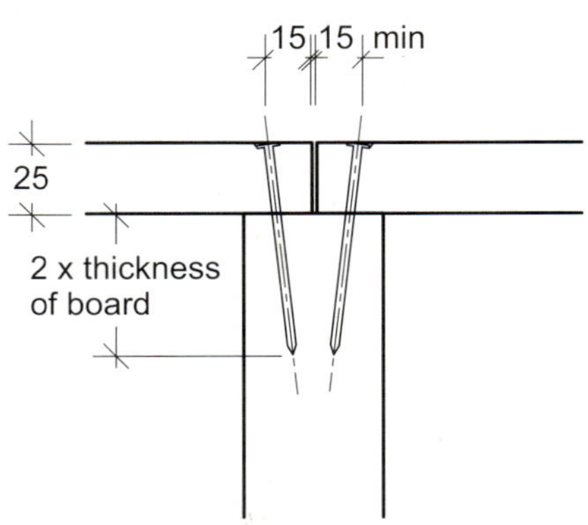

Nail fixings at butt joints between boards min width 38 mm joist pre-drilled holes angled as shown annular ring shank nails

Figure 5.7 Nail fixing

5.3 Fixing and fixings

Boards can be fixed to joists by various means – nailing, screwing or proprietary clips. The advantage of deck clips is that they avoid fixings being visible on the top surface of the deck boards. They also eliminate the risk of injury that could result from a nail or screw head becoming proud of the deck surface due to timber movement. Disadvantages can be that insufficient gaps are provided between the boards, the clips may loosen if the boards shrink, and it may be difficult to remove individual boards if necessary. Deck clips vary considerably in design so the method of fixing and detailed information should be obtained from the manufacturer.

Austenitic stainless steel is the preferred choice for ring shank nails or screws, although zinc plated or epoxy coated fixings can also give a good service life. A disadvantage with plated fixings is that if used with a timber containing a high level of tannin, such as European oak or Western red cedar, damage to the coating during fixing can result in corrosion of mild steel and consequent staining of the wood.

For nailing or screwing the boards there are some basic principles of good practice to be followed. For boards above 75 mm width two fixings should always be used, and these are best placed at quarter points on the board width. This reduces the risk of edge splitting, but also limits the width between the fixings if there is any moisture movement in the board. For boards 75 mm wide or less, single fixings can be used but these will offer little resistance to the board distorting sideways between the supports, see Figure 5.5. Staggering the fixings alternately to either side of the centre line can provide some resistance to lateral distortion.

If the board surface is grooved the fixings should lie in the grooves to prevent shrinkage of the boards causing the fixings to protrude and become a tripping hazard. The heads of nails or screws should be set flush with the base of the groove so that they do not block the flow of water, see Figure 5.6.

Nailing is usually adequate for fixing softwood deck boards but annular ring shank nails should be used to provide adequate resistance to pulling out or 'popping'. Nail penetration into the supports for ring shank nails should be approximately twice the thickness of the board. If nailed at quarter points in the board width (assuming a maximum of 150 mm width) at 20% or less moisture content, any movement due to moisture variation, will be taken up by embedment in the board and possible flexing of the nails. If the boards are to have a plain unribbed surface, nails should be flush with the top surface of the board, but if the boards are at a high moisture content when laid, or heavy wear is expected on the deck, there may be some advantage in driving the nail slightly below the surface. Nails should not be closer than 15 mm to the ends of boards and it may be necessary to predrill the holes to avoid the ends of the boards splitting, see Figure 5.7. Ring shank nails should not be used if it may be necessary to remove boards in the future as this will only be possible by destroying the board.

For more heavily used decks the boards should be screwed to supports with fixings at quarter points in the board width, see Figure 5.8. Screws should preferably be stainless steel parallel sided (chipboard or special deck screws) with countersunk heads inserted into pre-drilled holes. Because a dense hardwood will not compress as easily as a softwood, additional clearance should be provided between the screw shank and the board by pre-drilling the holes 2 mm oversize to allow for any moisture generated movement. This clearance should accommodate possible shrinkage of a board up to 150 mm width, if this is laid at not more than 20% moisture content. The countersunk heads of the screws will hold the board down adequately despite the

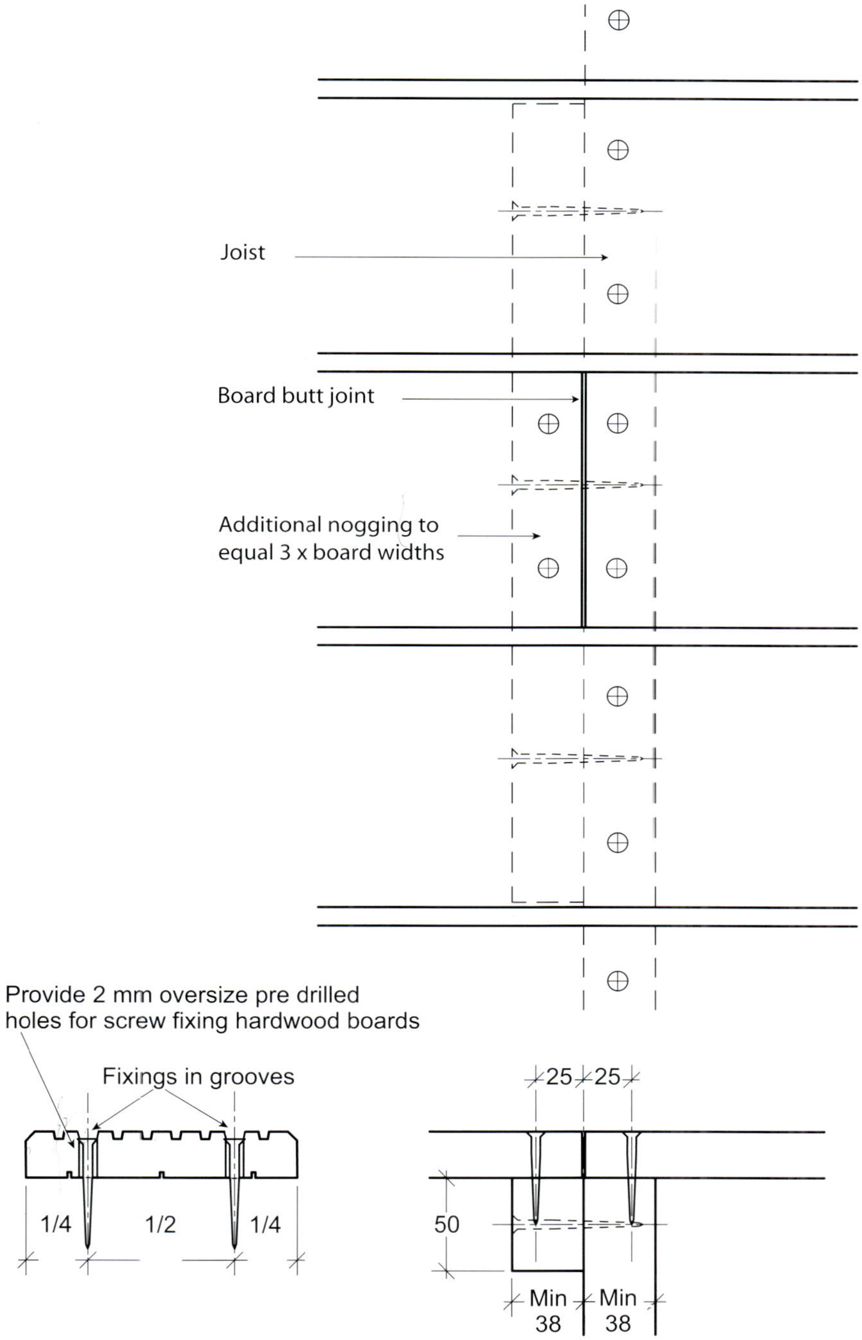

Screw fixings at butt joints - double joists or additional noggings (min 38 mm) where butt joints occur - batten length 3 x board width min 2 x screw fixings or 3 x annular ring shank nails (see plan view above)

Figure 5.8 Screw fixing

clearance around the screw shank. The screw gauge must be sufficiently thick to ensure that the countersunk head overlaps the pre-drilled hole sufficiently.

If a larger hole is necessary to take up the additional shrinkage of a wider board, or of an 'undried' board, it may be necessary to add washers under the screw head in recesses in the board. These washers should be of stainless steel with either a hole larger than the screw shank, or a recess in the board that will allow for sufficient movement of the board without the washers binding on the edges of the recess. In this case, round or pan headed screws may be more appropriate than countersunk.

Screws have the advantage that if there is shrinkage in the board thickness they can be more easily tightened than nails and boards can be easily removed in the future. Countersunk screw heads should be slightly recessed below the wearing surface of plain boards, but not to the extent that the holes can accumulate standing water. Fixing screws should preferably be kept at least 25 mm from the ends of boards to avoid the risk of end splitting. If boards are butted end to end this may mean that they cannot be fixed to a single joist.

Softwood boards are usually limited to a maximum length of about 4.8 m and generally any deck longer than this will require the boards to be butt jointed. Hardwood deck boards may only be available in shorter lengths. This should be considered in the board layout and for economy the boards should be in lengths that are increments of the chosen joist spacing. If the end joints are to be staggered they should preferably occur on the same line and regular board lengths should be used, rather than random lengths. This is particularly important if joists need to be doubled up in some places to provide sufficient fixing for the board ends, otherwise the number of joists may be unnecessarily increased. If boards are to be laid so that there is no obvious pattern to the joints, it may be necessary to add short lengths of batten to the sides of the joists coincident with the ends of the boards in order to provide sufficient space for fixing the boards.

Figure 5.9 Diagonal boards laid with two fixings per board

Boards laid diagonally are occasionally butted together on a single joist, with a single fixing in the corner of each board but this can lead to the corners of the board lifting and causing a tripping hazard and is not good practice. To avoid this alternate boards should overlap the width of the joist sufficiently to permit two fixings, see Figure 5.9. If boards are to be laid diagonally the larger span for the board between joists should be taken into account when choosing the thickness of the board. Boards laid diagonally have the advantage that they can provide good lateral stiffness to the deck structure.

A possible alternative method of fixing hardwood decks is to screw battens to the underside of the boards (allowing an oversize clearance hole in the batten) to form a panel. The panel can then be reversed, laid on the substructure, and the battens fixed to the substructure by screwing through the gaps between boards, see Figure 5.10. Besides eliminating all surface fixings, it is a useful way to either fix a timber deck to a steel substructure, or to provide panels that can be removed for cleaning purposes. This method will give a 'panelled' appearance to the deck as the ends of all the boards will coincide.

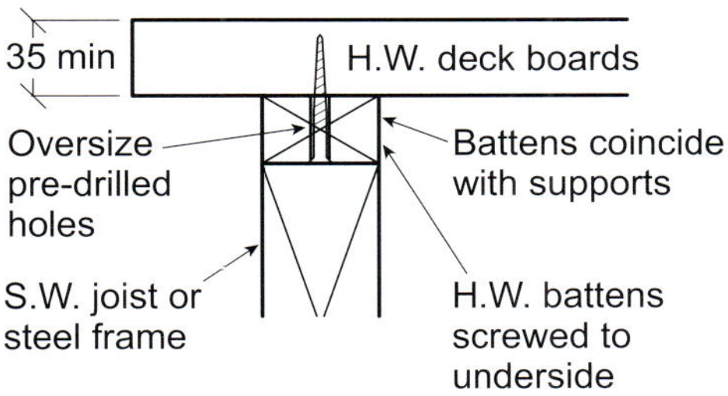

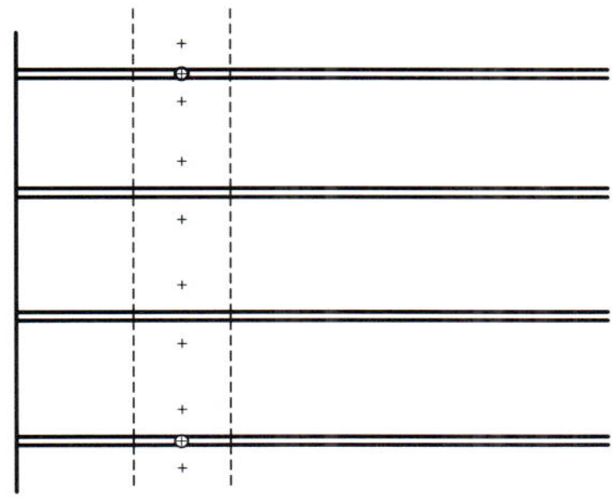

Figure 5.10 Back fixing boards

Where decking boards are cut around, or abut against posts, a clearance of 6 -10 mm should be provided so that any water running down the post will not be absorbed into the end grain of the boards. Providing sufficient clearance gaps is particularly important if the deck boards or other parts are to be finished with stain as it allows space for brush coating the ends or edges of the members when maintenance is required.

5.4 Fixing boards to a steel substructure

Where boards need to be fixed directly to steel, this would normally be by bolts through pre-drilled holes or with self-drilling screws.

For exposed decks stainless steel is the preferred material for fixings. However stainless steel is not compatible with mild steel because, in the presence of moisture, electrolytic action will take place between the two metals. This limits the choice of fixing so only galvanised or epoxy coated bolts, rather than self-drilling fixings can be used. Where the boards bear directly on steel members moisture can also be trapped between the steel and timber, leading to the risk of long term decay in the timber and corrosion in the metal. If this form of construction is essential the boards should be held off the steel by large plastic washers or packers that allow air circulation and drainage between the materials. With a steel substructure it is preferable to introduce treated or naturally durable wood battens between the boards and the steelwork. These battens can be attached with galvanised bolts or self drilling screws to the steel framing, but the boards can then be fixed to the battens with stainless steel wood screws or nails. This has the advantage of reducing the number of fixings directly into the steel which will also reduce cost. See Figure 5.11.

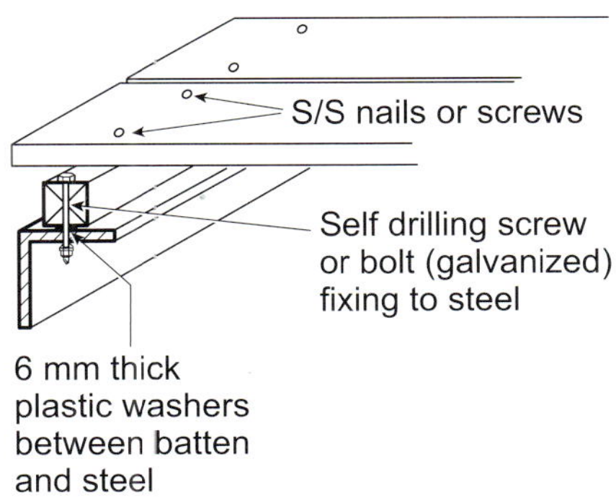

Figure 5.11 Fixing boards over steel frame

6 Superstructure

Ancillary components to timber decking such as guard rails, supports, balusters, stairs etc should be sufficiently robust to take both predicted loading and possibly more unpredictable loads such as people crowding against, or climbing on handrails, heavy objects or bikes being crashed against balusters etc. The possible occurrence of these additional loads should be considered in the initial structural design of the deck and substructure.

If a change in level of 600 mm or more occurs within the deck area, or between the deck and the surrounding ground level, a guard rail should be provided. Guard rails and balusters should conform with the recommendations in the relevant Building Regulations which prescribe their heights. These should be not less than 900 mm high for a low level deck but increased to a minimum of 1100 mm high for a deck more than 600 mm above ground level. Horizontal guard rails or vertical balusters should not be more than 100 mm apart. A guard rail for a low level residential deck should be designed to resist a maximum lateral force of 0.36 kN/m and for a high level deck 0.74 kN/m. Staircase pitch and slopes of ramps etc are also defined in the Regulations.

Figure 6.1 Guard rail surrounding high level deck
Photo: Richard Burbidge

6.1 Design detailing

Besides having sufficient strength to take these loads, all the component parts must be as durable as the decking and substructure. This durability should be achieved primarily by good detailing, such as protection of exposed end grain from contact with wet surfaces or surface water, avoiding any positions where water can be trapped, and providing sufficient clearance at joints to allow drainage and ventilation around as many faces of wood members as possible.

Water should be shed from any horizontal surfaces by sloping the top of the section, and in particular the tops of vertical members where the end grain is exposed. Where possible these ends should be capped by horizontal members, such as hand rails, or be separately covered with a wood capping piece or metal flashing. See Figures 6.2 - 6.3.

Posts supporting guard rails are often extensions of the main supporting posts, but as these may be widely spaced, it is usual to introduce intermediate posts to reduce the span of the guard rails or handrails. These posts are normally bolted into the face of the perimeter beams and support horizontal rails to which the balusters are fixed. Where the posts are bolted to the perimeter beam, the connections must be adequate to take the lateral forces on the guard rail.

Supports can either be rectangular posts or round poles. Connections may be more difficult with round poles, as they need to be notched to provide flat surfaces for connections but they have the advantage that any non orthogonal changes of direction to follow the shape of the deck can be more easily accommodated.

To provide sufficient stiffness, posts should not be less than a nominal 100 x 100 mm square and poles not less than a nominal 150 mm diameter to allow for notching. These posts or poles should be strong enough to transfer forces from the guard rail to the perimeter beams. Posts or poles of this size or bigger will appear more robust than smaller sections which can appear 'flimsy' when used externally. Children climbing guard rails should be discouraged and vertical balusters provide less opportunity than horizontal members, which are not recommended for high level decks, see Figure 6.4.

There is a particular requirement for handrails and guard rails in that they should be splinter-free. Care should be taken to ensure softwood handrails are of appropriate quality. Alternatively it may be preferable to use a hardwood rather than a softwood for these members.

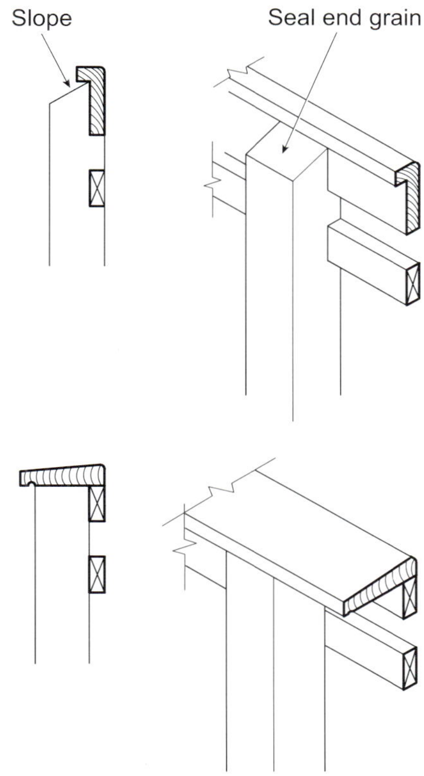

Figure 6.2 Handrail details

Figure 6.3 Handrail and baluster details

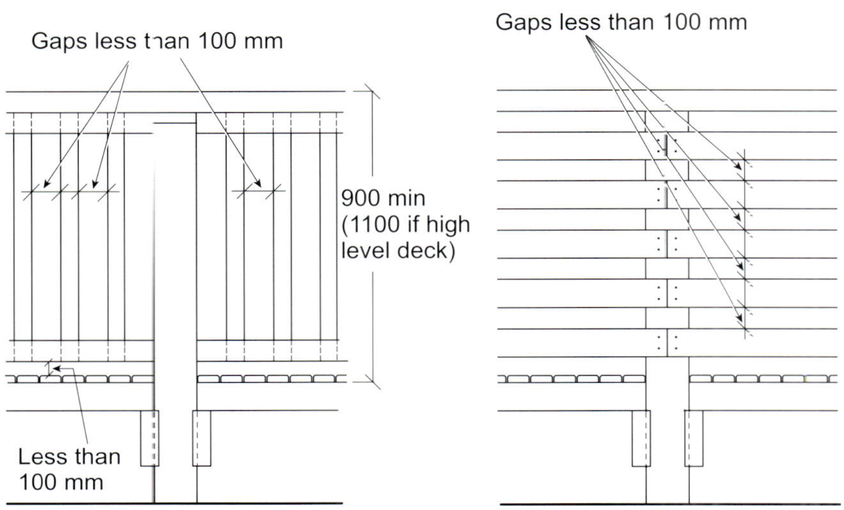

Typical baluster detail
Baluster on deck side of rail

Typical horizontal rails; not suitable for high level decks
Rails can be either side of posts

Figure 6.4 Baluster design

Figure 6.5 Balusters and handrail protecting a raised deck. Protective seating is used as feature at the lower level
Photo: HLD Ltd

If pergolas or screens are also to be part of the deck design, the simplest form of construction is to extend the deck supports to a sufficient height to support them, but it may be necessary to increase the size of the post to ensure sufficient stiffness and strength for the additional height. The tops of the posts are usually connected with edge beams which support the pergola and possibly the screen framing. It should be remembered that as foliage, such as creepers or vines, grows over a pergola or screen, increased wind and loads will be generated on all the structural members. This should be taken into account in sizing the sections and in designing the joints between members. In some cases it may be necessary to add some form of wind bracing to the whole structure. See Figures 6.6 and 6.7.

6.2 Superstructure fixings

All bolts, nuts, coach screws and washers used in jointing the ancillary components should be of hot-dipped galvanised mild steel or stainless steel to avoid corrosion. Nails are not appropriate for fixing any part of a superstructure, and fixings should be limited to screws, coach screws or bolts. For some highly loaded connections, it may be advisable to use metal angle cleats rather than larger screws or bolts.

It is beyond the scope of this manual to provide all the details necessary to construct these ancillary components, but the diagrams show some typical details. These may provide adequate guidance for the design of ancillary components to domestic decks, but for decks for public use it would be advisable to employ a specialist deck designer to size the components and design the necessary connections to ensure sufficient strength and stability.

The Timber Decking Association publishes a booklet *Deck parapet design and construction*, covering the structural and detail design of handrails, newel, posts, balusters etc where compliance with the Building Regulations is required.

BM TRADA Certification runs a Q-Mark scheme for timber balustrades, providing assurance that the product has demonstrated the required performance levels through independent accredited testing or appraisal and that every product manufactured offers the same level of performance through an approved quality management system. See www.bmtrada.com for more information.

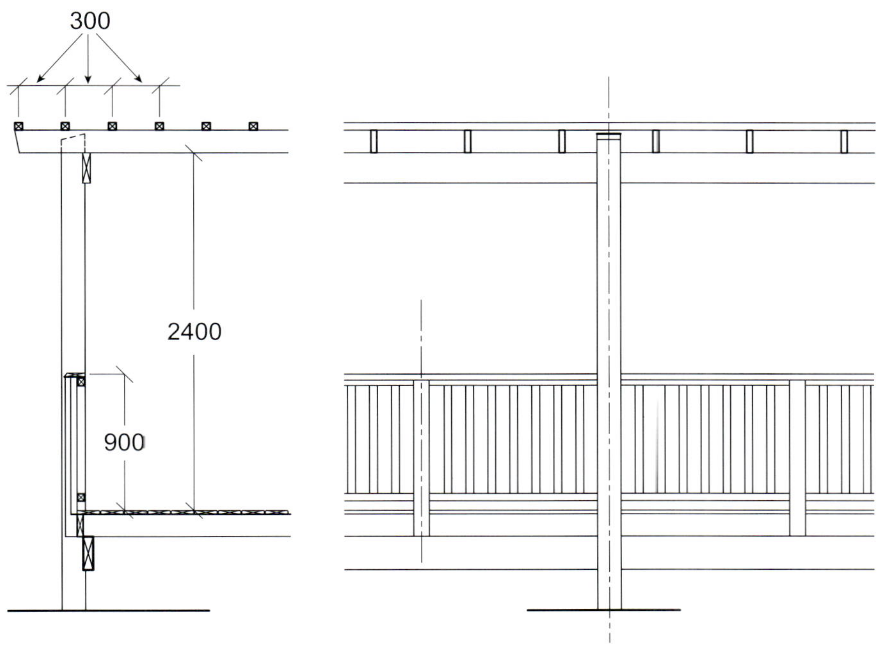

Posts carried through to support hand rail and pergola

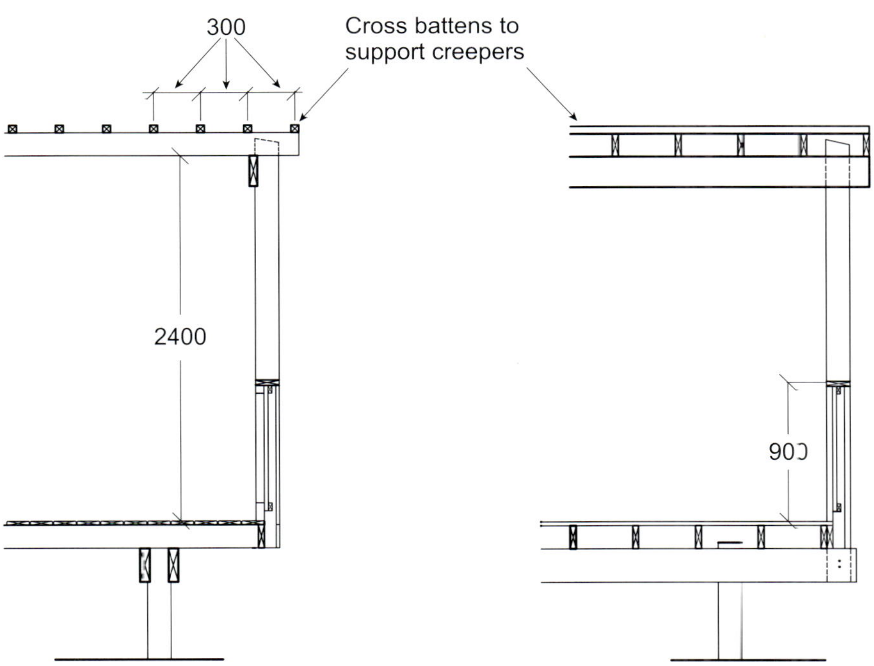

Joists and beams cantilevered to support hand rail and pergola

Figure 6.6 Standard pergola support

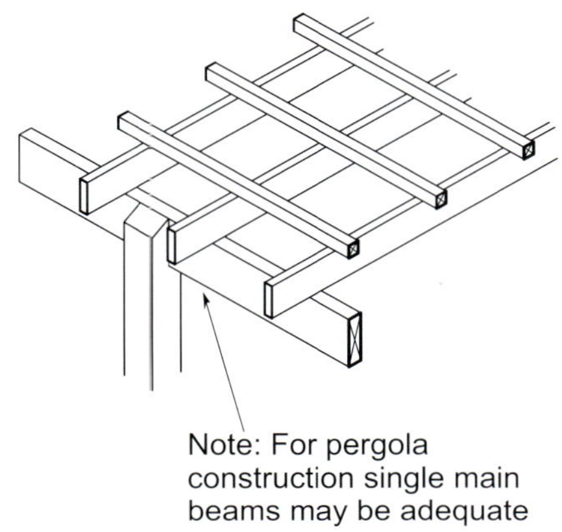

Figure 6.7 Pergola detail

7 Construction

7.1 Storage of materials

When constructing a deck it is important that all the component parts and materials are in the best condition and in particular at a moisture content close to the level likely in use. In the British Isles the basic moisture content of exposed timber is likely to be about 18% and this may vary during the course of the year by about 2% either way.

Exposure to direct rainfall may raise the surface moisture content above this level and exposure to hot sun may bring the surface moisture content well below, but it is the moisture content in the core of the wood that is most critical because this will determine the amount of swelling or shrinkage of the section that will take place, related to the size of the section. For this reason it is preferable to use smaller structural sections, doubling up if necessary, so that any moisture movement in the long term does not lead to shrinkage that can loosen bolted or screwed connections unduly.

It is important that precautions are taken to ensure that wood that has been dried to the specified level does not pick up moisture on site during storage. All sections should be stacked on bearers well clear of any wetted surfaces, preferably at least 150 mm above ground level. All structural members, ie posts, beams and joists, should be stacked in such a way that air can circulate around them, by including battens between each layer. Deck boards can be stacked in packs, and may be delivered already shrink wrapped in plastic, which should not be removed until the boards are ready to be laid. All unwrapped timber sections should be protected from rain by waterproof covers but these should not be wrapped so tightly that they prevent air circulation around the timber. The cover should be held clear of the stack by more battens to prevent any condensation forming under it wetting the top surface of the timber, Generally, the less time timber is stored on site the better in reducing the risk of any moisture pick-up.

Careful storage will also ensure that the wood does not get dirty or stained before construction of the deck. This is particularly important if the deck is to be left unfinished, as any marking may be difficult to remove or conceal. Water or oil-borne dirt may penetrate deeply into unfinished wood. Deck boards pre-finished with a moisture repellent treatment will be less susceptible to deep staining but can still be marked on the surface.

7.2 Constructing the deck

It is important that concrete foundations for the posts are in position and the concrete has gained sufficient strength before construction of the deck structure begins.

Preparatory groundwork, such as removing vegetation, applying weed killer, compacting the soil, forming falls for drainage etc, should also be complete. Laying polythene sheet below the deck at this stage will help protect the wood from dirt and moisture from the ground but could complicate fixing the posts to foundations and it might also be damaged by the construction process. Hence, if the polythene is laid it is better to also add the gravel ballast at this stage as protection.

Fixing posts to the top of concrete pads is easier if the pads project approximately 150 mm above ground level. This will also raise the timber above the risk of splashing and above any gravel ballast. Because of the difficulty of locating poured concrete foundations accurately and achieving a good appearance if they are visible above ground, it may be easier to use precast pads, precast concrete drain rings or specialist foundation tubes as formers. These can be accurately positioned and later filled with poured concrete to which steel supports can be bolted or in which the support posts can be embedded.

If it is intended to embed the support posts directly into the foundations they can be inserted into the precast concrete rings or specialist formers, accurately located, and then packed around with concrete containing no sand, or with gravel. The precast ring should be firmly located on a poured concrete base which will provide bearing for the base of the post. Provision should be made to allow water to drain through the base.

The construction of the deck should always start with any connection to an existing building to ensure that the construction is aligned and square to the face of the building, and at the correct level.

The level and squareness of the whole deck will also be very dependent on the accurate positioning of the supports. One way of ensuring the accurate location of a post is to provide some adjustment between the base and the post by providing oversize holes in the steel support bracket bolted to the column base Alternatively oversize holes can be cast in the concrete to allow some tolerance on the bolt positions when they are grouted in position, see Figure 4.5. After the posts are installed plumb and true they should be connected together by the edge beams. To provide a seating for these perimeter beams or any other internal beams, the posts are sometimes notched. This notching, if left until the posts are erected, can be adjusted in height and depth to correct any small inaccuracies in plumb, height or alignment of the posts.

Horizontal beams, or plates on supporting walls, will set the overall squareness and level of the whole deck construction. Beams are often double members bolted either side of posts, which provides an easy way of fixing, but they must be accurately levelled in relation to each other.

For simplicity in construction the joists should be offset from the beams and post positions so that they can be directly fixed to beams, rather than to the tops of posts

If the posts are to be carried through the deck to support guard rails, screens, pergolas etc, the joists will not then need to be cut back at the post. In this case it will be necessary to provide support for the ends of the notched deck boards. If double beams are used with joists set between, it will be necessary to include blockings between the beams coincident with joist centres to support any board laid along the centre of the beam. Joists should be accurately positioned over, or between the beams, ensuring that they are evenly spaced and perpendicular to beams.

The overall dimensions of the deck should be checked before the boards are laid so that a combination of full width boards and gaps between will cover

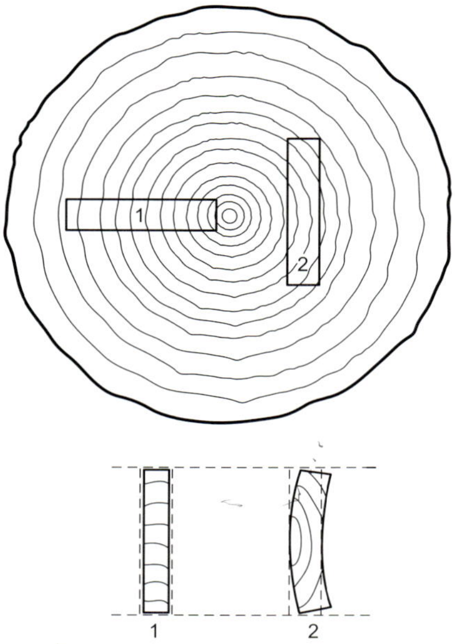

Figure 7.1 Shrinkage and distortion of boards after drying

the area, without the need to split boards. Adjustment can usually be made on the width of the gap between boards to achieve the dimension required, but this should preferably never be less than 5-6 mm and not more than 10 mm. Undried (green) boards should be laid with smaller gaps as shrinkage of the boards will cause the gaps to widen.

To ensure that the required gaps are maintained, and the boards aligned accurately perpendicular to the joists, temporary spacers should be inserted between the boards as they are laid.

Laying the deck is easier if the ends of the deck boards oversail the edge support. They can then be cut back to an accurate line after fixing.

The annual rings in a piece of timber (viewed end on) tend to flatten out as it dries. Therefore if the ring orientation is obvious in a plain board it is preferable to lay it 'heartside' up so that if there is any distortion of the board it will 'bow' and shed water, rather than 'cupping' which will restrict drainage. However, with decks laid at the correct moisture content and restrained by adequate fixings, cupping or bowing is rarely a problem, particularly if the boards are radially sawn.

7.3 Temporary protection

Once the deck is laid it is advisable to protect the surface during the construction of any of the ancillary parts. Alternatively the deck boards can be left off until everything else is built, but the structure will then be less stable and would not provide a working platform.

A sheet of polythene is sometimes used to protect the boards from surface moisture after they are fixed but condensation can sometimes form underneath this in sunny conditions, wetting the boards. Polythene will also provide little protection against impact or scoring. A preferred method is to cover the boards temporarily with sheets of hardboard, OSB, or plywood to protect against accidental impact, dirt becoming embedded or spillage.

If the deck is to be left unfinished it is particularly important to protect it against staining during the construction process but temporary protection, particularly polythene, can also possibly result in other forms of stain. If there is a sustained moisture level under the cover for instance, mildew or blue stain may occur in the boards which is then difficult to clean off.

8 Finishing and maintenance

8.1 Finishes for new decks

Although dense woods generally have a natural resistance to wetting, low density softwoods are quite permeable and will therefore pick up moisture easily. This can lead to alternate swelling and shrinking of the wood which can result in surface checking and possibly splitting.

To increase the water-repellency of the timber, wax emulsion can be added as part of the preservative treatment. Alternatively there are a number of wax, stearate or silicone clear penetrating finishes available for site application. These treatments do not affect the appearance of the wood, or prevent surface bleaching, but make the wood more resistant to picking up moisture. These treatments will generally need regular re-coating to maintain their moisture resistance. Traditional coatings such as teak oil tend to oxidise too rapidly and will not therefore provide effective protection against moisture penetration, unless frequently re-applied. Unmodified linseed oil is not recommended because it encourages the growth of surface mould, but oil finishes are available which waterproof, protect and resist surface mould growth.

Figure 8.1 Deck boards freshly treated to demonstrate water repellent treatment
Photo: HLD Ltd

Clear coatings, such as varnishes, that form surface films will tend to yellow when exposed to ultra violet light. Eventually they may flake off because they are also not flexible enough to take up any moisture movement of the wood. They have limited vapour permeability so any moisture that penetrates the surface may be trapped in the wood and could add to the risk of flaking. These types of finish are not recommended for use on decks.

8.1.1 Stains

Modern decking stain finishes should always be used, rather than finishes such as varnish, because they penetrate the surface, are sufficiently flexible to take up movement of the wood without cracking, yet tough enough to resist abrasion due to foot traffic. They are also vapour permeable so that any moisture that does penetrate into the wood can dry out again through the surface finish. Areas where the surface coating becomes worn will also be less obvious if there is some surface penetration of the colour. Stains can be either translucent or semi-translucent, showing the grain of the wood to varying degrees. They are pigmented and can alter the colour of the wood.

To resist the effects of sun, and thus to prevent wood bleaching, coatings need to contain some pigment that will filter out the ultra violet light. Pigmented finishes can either simulate the original colour of any wood, or be of any other colour, but it is preferable not to use significantly darker tones because these will result in a higher temperature and cause increased surface checking.

Treatment with stains can be preceded by the application of a tinted base coat. This base colour, even when not protected by an additional finish, will last for several months and if visually acceptable the boards can generally be left unfinished, as the treatment provides improved resistance to ultra-violet degradation of the surface.

Figure 8.2 Coloured stains
Photos: TDA

For softwoods generally where there is a high risk of regular wetting, eg pool sides or under heavy foliage, there are advantages in reducing the uptake of moisture by the use of clear penetrating seals to increase the water repellency of the timber whilst also allowing it to 'breathe'.

A coating on a sawn, rather than a planed surface, will tend to last longer before requiring maintenance due to its increased uptake. Moisture-resistant treatments or coatings to the decks may not be compatible with water-based stain finishes applied subsequently because they will inhibit the penetration of the stain and result in patchiness and reduced life. Organic solvent-based stains may be suitable for application over moisture-resistant treatments but the stain manufacturer should be consulted. Alternatively stains may be applied at a later date after the moisture resistance at the surface of the board has diminished through weathering.

Conventional stain finishes can be applied to parts of the deck such as guard rails, balusters and other surfaces and will last well because they are not exposed to pedestrian wear. It is not necessary to apply finishes to a substructure except for visual reasons, for instance under a high level deck.

Before applying the stain, the surface of the wood should be brushed clean and any underlying discolouration or marks removed. The finish should only be applied within the range of weather conditions and temperature suggested by the manufacturer.

If a surface stain finish is to be applied it is preferable to coat the timber either before erection or as soon as the construction is sufficiently complete. A finish not only provides some protection against wetting and staining, but also protects the wood from ultra-violet light, which can affect the surface of the wood and reduce the bond of any subsequently applied finish.

Usually a low-build stain, possibly slightly diluted, is used for the first coat to ensure good penetration into the wood. Further coats may be required according to manufacturers' recommendations. It is important to ensure full, even coverage of the deck boards to provide the maximum resistance to wear but it is also important not to overload the stain, otherwise 'breathability' may be compromised. All end grain should be sealed before the component is installed, either with a liberal application of a proprietary sealer or several coats of stain.

Where a deck is left unfinished and a moisture-resistant treatment is not specified, there is some advantage in sealing the end grain of boards with two or three coats of clear or lightly pigmented stain, in addition to any brush-applied end grain preservative. This reduces moisture pick-up and consequent swelling and shrinkage of the boards which can lead to end splits. Stains on vertical surfaces dry relatively quickly and surface contact is unlikely to cause damage. On decks the finishes must be allowed to fully cure before there is any traffic on the surface.

All surface finishes need regular inspection and possible re-coating and this may be more frequent on decks subjected to high levels of surface abrasion and erosion from weather.

8.1.2 Anti-slip finishes

Coatings are available to improve the slip resistance of decks. These are either clear coatings incorporating micro-beads, grit aggregate, or grit-faced adhesive strips that add a textured finish to the wood. Any stain finish must be applied before the coating or strips because they will prevent penetration of the stain. As a clear finish they will have limited life but this might be extended for those that include a coloured pigment.

8.2 Maintenance

All timber deck structures should be designed to ensure structural integrity and to give a long-term life expectancy. Some suppliers offer standard guarantees on workmanship and extended guarantees on materials. However it is important to realise that routine care and maintenance can also enhance the long-term appearance and durability of the deck structure.

As timber weathers, surface checks and splits are inevitable and result from the natural variation in moisture content and the effects of intermittent wetting and drying of the timber surfaces. Such features should not be considered as defects and have no structural effect on the performance of the deck structure. However, if major cracks or through-splits occur, it is advisable to contact the deck supplier or contractor immediately to avoid structural failure or consequential damage.

The regular application of surface treatments and finishes can reduce the uptake of moisture and the consequent swelling and shrinking that leads to checking but these products should always be used and maintained strictly in accordance with manufacturer's recommendations. Where decks are under trees or in close proximity to foliage, occasional, but regular, applications of an anti-fungicide solution will prevent accumulation of mildew and algal growth.

8.2.1 Cleaning

Any decking will benefit from regular cleaning although the frequency will depend on usage.

Occasional brushing with a stiff broom will normally be adequate for most decks to remove dirt, dead leaves or any slight surface growth in areas where there may have been little wear. The brush should be stiff enough to remove dirt that may have accumulated in drainage grooves or gaps but not so stiff as to damage any applied surface finish.

Brushing the deck regularly will not only maintain the appearance but will reduce the risk of slipperiness developing from surface growth, ingrained dirt, or standing water accumulating due to blocked grooves or gaps between the boards.

Cleaning public decks is frequently carried out by mechanical brushes but rotary brushes are less effective than the cylindrical type in cleaning dirt from grooves and gaps. One advantage of running boards along the deck, rather than across, is that the grooves and gaps can be more easily cleaned by machine. Coarse fibre brushes will tend to remove a surface finish very quickly and can seriously scour the surface even of dense hardwoods. It is important that an appropriate brush stiffness is chosen. Steel brushes are not recommended because fragments or swarf may become embedded in the timber and corrode, causing staining, particularly in timbers with a high level of tannin, such as oak. Alternatively, the use of specific timber deck cleaners will remove growths, clean and may brighten greyed timber.

On large decks for public use it is especially important to keep the drainage grooves clear because, even with clean drainage grooves, it takes longer for any surface water to drain away from a large area. If a non-slip surface finish is applied, or there are non-slip inserts in the grooves, particular care must be taken with mechanical cleaning to ensure that the surface finish is not removed or the inserts dislodged by the brushing.

It is not advisable to wash down timber decks with large quantities of water or hoses because there is a risk of wetting the boards sufficiently to cause moisture movement. This could affect the life of any surface finish and lead to increased surface checking when the boards dry in the sun and shrink. However cleaning with a high pressure hose which combines air with a small quantity of water, will dislodge dirt effectively without excessively wetting the wood.

Local soiling and surface growth can usually be removed by mild detergent solutions but if stains are more deep-seated on unfinished boards it may be necessary to use specialist deck cleaners which remove dirt, algae and mildew. Dilute household bleach can be used but the effect should be tested to determine the appropriate strength before it is used in any quantity on the boards. If the stains are from chemicals or metals and still resist this treatment it may be necessary to use special products available for deck restoration which can generally remove quite deep seated stains.

To avoid uneven bleaching, or variation in the colour of a stain finish, it is advisable occasionally to move around furniture, or other objects, kept permanently on the deck. Planters or plant pots should be stood on trays so that watering does not result in the deck under the pots being permanently wetted, leading to discolouration and mould growth. If the drip trays are supported on battens the top surface of the deck will also be well ventilated.

8.2.2 Tightening fixings and re-surfacing

After the deck has been in use for some time it may be necessary to carry out two forms of maintenance.

The first is to ensure that all the boards are firmly held down. Shrinkage and sometimes wear can result in the loosening or 'popping' of fixings. Nails should be driven home flush to the surface of the boards, and screws tightened as necessary. Occasionally it may be necessary to replace individual boards that have distorted or become excessively worn in heavily used areas of the deck.

Besides tightening the deck fixings it is advisable to also inspect any bolted or screwed connections in the sub-structure which may have become loose through shrinkage of the wood, and tighten these as necessary. It is advisable to first do this check about six months after the deck has been completed, by which time any substantial shrinkage in the wood should have taken place.

The second form of maintenance would normally only be required for unfinished decks where the surface may have become encrusted or worn over time. If the decks are of unfinished hardwood occasional mechanical scouring with stiff brushes or sanding will remove mould growth, loose material, ingrained dirt or surface staining but it is important to avoid scouring or sanding to the extent that the surface is excessively worn down or, alternatively polished to a level that would increase the slipperiness of the surface.

8.2.3 Re-finishing

Surface finishes will require regular inspection and occasional re-coating. Clear moisture resistant coatings should be re-applied in line with the manufacturer's recommendations, normally every two years. Pigmented stains will last several years (depending on the product, the quality of application, the exposure and usage of the deck). It is important that the deck and the grooves and gaps between boards are completely cleaned and allowed to dry thoroughly before the application of a new coat of stain.

If the existing stain has broken down to the extent that the wood has begun to turn grey in places, it is also important to brush away any loose fibres from the surface or treat with a cleaner before re-applying stain. However, it should be noted that if surface degradation has proceeded to the extent that greying has occurred, this is usually sufficient to compromise the performance of any future coatings or treatment.

The deck should be as dry as possible before the stain is applied and the temperature should be within the range prescribed by the manufacturers. It is always advisable to use the same product for re-finishing to ensure that the new coating is compatible with the old.

References

British Standards

BS 1186 – 3: 1990 Timber for and workmanship in joinery. Specification for wood trim and its fixing.

BS 1282: 1999 Guide to the choice, use and application of wood preservatives.

BS 4978: 1996 Specification for visual strength grading of softwood.

BS 5268-2: 2002 Structural use of timber. Code of Practice for permissible stress design, materials and workmanship.
BS 5268-5: 1989. Structural use of timber. Code of Practice for preservative treatment of structural timber. (Obsolescent).
BS 5268-7.1: 1989 Structural use of timber. Recommendations for the calculation basis for span tables. Domestic floor joists

BS 5589: 1989 Code of Practice for preservation of timber. (Obsolescent).

BS 5756:1997 Specification for visual strength grading of hardwood.

BS 6399-1: 1996 Loading for buildings. Code of Practice for dead and imposed loads.

BS 8000-5: 1990 Workmanship on building sites. Code of Practice for carpentry, joinery and general fixings.

BS 8417: 2003 Preservation of timber. Recommendations.

BS EN 335-1: 1992 Hazard classes of wood and wood based products against biological attack. Classification of hazard classes.
BS EN 335-2:2006 Durability of wood and wood-based products. Definition of use classes. Application to solid wood

BS EN 336:1995 Structural timber. Coniferous and poplar. Sizes. Permissible deviations.

BS EN 338: 2003 Structural timber. Strength classes.

BS EN 350-1: 1994 Durability of wood and wood based products. Natural durability of solid wood. Guide to the principles of testing and classification of natural durability of wood.
BS EN 350-2: 1994 Durability of wood and wood based products. Natural durability of solid wood. Guide to the natural durability and treatability of selected wood species of importance in Europe.

BS EN 460: 1994 Durability of wood and wood based products. Natural durability of solid wood. Guide to the durability requirements of wood to be used in hazard classes.

BS EN 599-1: 1997 Durability of wood and wood-based products. Performance of preservatives as determined by biological tests. Specification according to hazard class.

BS EN 912: 2000 Timber fasteners. Specifications for connections for timber.

BS EN 942: 1996 Timber in joinery. General classification of timber quality.

BS EN 1313-1: 1997 Round and sawn timber - permitted deviations and sawn sizes. Softwood sawn timber.

BS EN 1313-2: 1999 Round and sawn timber. Permitted deviations and preferred sizes. Hardwood sawn timber.

BS EN 1912: 2004 Structural timber. Strength classes. Assignment of visual grades and species

BS EN 14081-1: 2005 Timber structures. Strength graded structural timber with rectangular cross section. General requirements.
BS EN 14081-2: 2005 Timber structures. Strength graded structural timber with rectangular cross section. Machine grading. Additional requirements for initial type testing.
BS EN 14081-3: 2005 Timber structures. Strength graded structural timber with rectangular cross section. Machine grading. Additional requirements for factory production control.
BS EN 14081-4: 2005 Timber structures. Strength graded structural timber with rectangular cross section. Machine grading. Grading machine settings for machine controlled systems.

BS EN ISO 1461: 1999 Hot dip galvanized coatings on fabricated iron and steel articles. Specifications and test methods.

BS EN 1995-1-1: 2004 Eurocode 5: Design of timber structures. General. Common rules and rules for buildings.

BS EN ISO 3506-1: 1998 Mechanical properties of corrosion-resistant stainless-steel fasteners. Bolts, screws and studs

BS EN ISO 3506-2: 1998 Mechanical properties of corrosion-resistant stainless-steel fasteners. Nuts

BS EN 13556: 2003 Round and sawn timber. Nomenclature of timbers used in Europe

Regulations

Department of Finance and Personnel. Fire safety. Technical Booklet E: 1994.
Department of Finance and Personnel. Stairs, ramps, guarding and protection from impact. Technical Booklet H: 2000.
Department of Finance and Personnel. Access facilities for disabled people. Technical Booklet R: 2000.

Office of the Deputy Prime Minister. The Building Regulations 2000. Fire safety. Approved Document B. 2000 edition consolidated with 2000 and 2002 amendments. TSO. 2004..

Office of the Deputy Prime Minister. The Building Regulations 2000. Access to and use of buildings. Approved Document M. TSO. 2004.
Office of the Deputy Prime Minister. The Building Regulations 2000. Protection from falling, collision and impact. Approved Document K. 1998, amended 2002. TSO 2006.

The Scottish Building Standards. Technical Handbooks. Astron. Edinburgh. 2004.

TRADA Technology publications

for more information go to www.trada.co.uk

British grown hardwoods - The designer's handbook. 1996.

Wood Information Sheets

1 - 17 Structural use of hardwoods.
1-25 Structural use of timber: An introduction to BS 5268-2: 2002.
1 - 31 Timber for landscape architecture.
1-37 Eurocode 5 – An introduction.
2/3-1 Finishes for external timber.
2/3-10 Timbers - their properties and uses.
2/3-16 Preservative treatment for timber - a guide to specification.
2/3-32 Timber: fungi and insect pests.
2/3-33 Wood preservation - Chemicals and processes.
2/3-37 Softwood sizes.
2/3-51 Timber engineering hardware.
2/3-52 Fasteners for structural timber: Nails, screws, bolts and dowels.
2/3-54 Exterior coatings on 'alternative' hardwoods.
4-7 Timber strength grading and strength classes.
4-12 Care of timber and wood-based products on building sites.
4-14 Moisture in timber.
4-28 Durability by design.

Timber Decking Association publications

for more information go to www.tda.org.uk

TB02 Statutory Regulations.
TB04 Deck parapet design and construction.
TB08 Metal fixings.

Wood Protection Association publications

for more information go to www.wood-protection.org

Industrial wood preservation: Specification and practice. Derby. 2006.

Appendix I Timbers for decking

Chapter 3 'Performance' set out the functional requirements for a deck and considered the properties required of a timber to be used in decking. These technical criteria should be used in assessing the suitability of a timber species, along with other considerations such as sustainability, cost and availability.

AI.1 Sustainability and certification

A significant and growing proportion of the world's forests are well-managed and both national governments and the timber trade want to encourage this – and to improve the situation in those countries where forest management policies are weak. To this end, third party certification schemes have been developed to allow well-managed forests to be identified and the products being exported from them to be traced through the supply chain.

All specifiers and users of timber and wood-based products are encouraged to obtain their supplies from certificated sources. Government policy requires all departments and their agencies to actively seek to buy timber and paper products from legal and sustainable sources. The policy is not mandatory for local authorities but they are encouraged to adopt it, or something similar. The Central Point of Expertise in Timber (CPET) is the government's own centre for developing and implementing its Sustainable Timber Procurement Policy.

It is important that dialogue is established with potential suppliers at an early stage in the procurement process. Suppliers should have a thorough knowledge of the origins of their stocks and be prepared to share this knowledge, supported by appropriate certificates, with their customers.

Third party certification in the timber sector is a two-tier process. At the first level, forests are independently certified to a recognised standard (Forest Certification). At the second level, operations in the timber supply chain are certified (Chain of Custody certification) to allow them to buy and sell timber from these forests as certified.

Certification can be performed against a number of different standards, the best-known of which are the Forest Stewardship Council (FSC) and the Programme for Endorsement of Forest Certification (PEFC). Other certification schemes often relate to timber production and supply in particular countries, such as the Canadian Standards Association (CSA), the Malaysian Timber Certification Council (MTCC) and the (US) Sustainable Forestry Initiative (SFI). These five certification schemes were evaluated (in 2005) by CPET as satisfying the requirements for either sustainable and/or legal timber.

The certification process is undertaken by an independent, suitably accredited Certification Body, such as BM TRADA Certification, amongst others.

Full certification of forest management and/or chain of custody to a recognised standard, such as those listed above, can be a very considerable undertaking in financial and organisational terms. For this reason, such certification is far more common in the forests and supply chains of the developed countries than in those of the developing nations.

However, where the forests of these developing nations are indeed being well-managed, it is important that their forests' products can be specified and used, even if full certification has not yet been achieved. The alternative is that their forest products will lose value and that there will be economic pressure to substitute their forests with some other land use.

Therefore Independent Verification has been developed to enable forest management policies in these countries and the emanating supply chain practices to be audited as sound and responsible. This can be a very important interim step in the often lengthy process of achieving full certification.

One such service, the Forest Products Chain of Custody Certification Scheme, is operated by BM TRADA Certification. It contributes to the further development of forest certification worldwide by lending support and encouragement to various non-certified initiatives that are progressing towards certification. This process, if given sufficient market incentive may, in due course, give rise to full forest management certification in countries such as Ghana and Cameroon who otherwise will find it increasingly difficult to sell timber products into the UK. It is in no one's interests if legitimate markets are closed to these countries because they cannot deliver full forest management certification.

All UK Timber Trade Federation members are required to adopt and comply with a Code of Conduct which includes a commitment to sourcing timber and timber products from legal and well-managed forests. Companies recognise that the independent certification of forests and the process chain is the most useful tool in providing assurances that the timber they deal in meets these requirements.

AI.2 Cost and availability

Most softwoods used for decking are available with third party environmental certification and manufactured decking components are readily available from a wide range of companies. The properties of softwood species commonly used in decking are given in the table in AI.3.

Hardwoods present a different picture. Those with well established properties and a history of use for external and structural purposes in the UK are included in the timber design code, BS 5268-2. Those which are suitable for decking are included in the table in AI.4. Environmental certification for these is variable depending on product and country of origin. In general terms, certified temperate hardwoods are more readily available than certified tropical hardwoods.

A number of lesser-known hardwoods with third party environmental certification, mostly from South America, are now available in the UK and are being used for decking. As the description implies, their properties are less well established and experience in their use in UK conditions is limited. Availability of these species may not be consistent so specifiers should check with suppliers at an early stage in the design of a project, particularly, if large quantities are required. Timbers in this category include angelim (*Hymenolobium* spp), cumaru (*Dipteryx* spp), itauba (*Mezilaurus* spp), ipê (*Tabebuia serratifolia*), massaranduba (*Manilkara bidentata*), pau amarelo (*Euxylophora paraensis*), and tatajuba (*Bagassa guianensis*).

A further aspect which needs to be considered in selecting timber for decking is the availability of particular dimensions. For example, while softwood decking may be available in lengths up to 5 m, tropical species may be generally available only up to 4 m long (although longer lengths may be available to special order). Temperate hardwoods may not be widely available in lengths exceeding 2.5 m, although longer lengths may be obtainable in small quantities. Similarly finished section sizes of hardwoods may differ from those of softwoods.

AI.3 Softwoods

Species/ Origin	Appearance	Durability	Strength	Moisture movement	Density (average) kg/m^3	Resistance to abrasion/ impact	Comments
Western red cedar Canada	Staight grained, few knots in most commercial supplies	Durable	Low	Small	390	Low - indents and wears; also tendency to split	Used extensively for decks in North America. Tannin content can affect ferrous metals.
British pine, European redwood	Clearly visible grain. Depending on grade can have large number of knots	Slightly durable. Requires preservative treatment	Medium (Imported stronger than home-grown)	Medium	510	Medium	Imported wood is closer grained than British timber. The timber contains resin, which can affect the performance of decorative finishes. May be left unfinished but surface treatment with water repellent will improve weathering.
Douglas fir British	Prominent grain - wavy	Slightly durable (requires preservative treatment)	Medium	Medium	530	Medium	Can be left unfinished. Tannin content can affect ferrous metals.
Douglas fir North American		Moderately durable	Medium (Imported stronger than home-grown)				Can be left unfinished. Tannin content can affect ferrous metals. Decks constructed from N American timber (heartwood) may not require preservative treatment providing the deck is correctly designed and detailed.
Larch British/ European/ Siberian	Prominent grain – straight (sometimes wavy) can be knotty	Moderately durable	Medium (Imported stronger than home-grown)	Medium	550	Medium/high	Can be left unfinished.
Southern pine USA	Clearly visible grain, few knots	Slightly durable (requires preservative treatment)	High	Medium	590	Medium/high (varies with density)	Used extensively for domestic decks in North America. The timber contains resin, which can affect the performance of decorative finishes.
Note: All durability rating refer only to heartwood - all timber containing sapwood should be treated with an appropriate preservative							

AI.4 Hardwoods included in BS 5268-2: 2002

Species	**Appearance**	**Durability**	**Strength**	**Moisture movement**	**Density (average) kg/m^3**	**Resistance to abrasion/ impact**	**Comments**
Balau (yellow)	Fine even texture	Very durable	Very high	Small	980	Very high	
Iroko	Coarse grain but not very visible	Very durable/ durable	High	Small	660	High	Care needed in specification and selection of timber to avoid problems of distortion.
Jarrah	Usually straight grain. Grain not strongly marked	Very durable	High	Medium	820	Very high	Very popular Australian timber for heavy duty decking.
Karri	Striped grain	Durable	Very high	Large	900	Very high	Popular Australian timber for heavy duty decking.
Oak, European	Straight to wavy grain. Broad rays	Durable	High	High	720	High	Tannin content can affect ferrous metals. Considerable tannin run-off if wood used undried. Can be left unfinished but subject to surface checking.
Opepe	Grain not noticeable but irregular coarse texture	Very durable	High	Small	750	Very high	Surface of timber develops numerous small checks when in service. These do not impair performance.
Teak	Sometimes figured usually straight	Very durable (Plantation-grown very durable/ moderately durable)	High	Small	660	High	Traditional timber for decking. Obtainable from a wide variety of sources including plantation-grown options.
Note: All durability rating refer only to heartwood - all timber containing sapwood should be treated with an appropriate preservative							

Appendix II Span tables

Span tables have been prepared to assist in the preliminary design of timber decks, eg for species selection and member sizing. However, the detailed structural design will need to be confirmed by an appropriately qualified engineer. TDA member companies provide this as part of their service.

To simplify selection and structural design, timber is grouped into a series of Strength Classes. Design stresses for these are given in BS 5268-2 Structural timber can be visually graded (to BS 4978 for softwoods or to BS 5756 for hardwoods) or machine graded (to BS EN 519 or to BS EN 14081, which will replace BS EN 519) direct to a strength class. Visually graded timber is allocated to the appropriate strength class in BS EN 1912 on the basis of its species and grade.

Span tables are presented for the species/ strength classes shown below. The species have been selected on the basis of their suitability for decking so other species should not be substituted for those of equivalent strength class.

Table AII.1 Timber species and grades meeting the strength classes in the span tables

Standard name	Marking code (from BS EN 14081 or BS EN 13556)	Visual strength grade
Softwoods:		
Strength class C16		
British spruce	WPCS	-
British pine	WPNN	-
British grown Douglas fir	PSMN	-
European redwood/whitewood	WPPA	GS
Strength class C18		
Western red cedar (imported)	THPL	SS
British spruce	PCAB	SS
British grown Douglas fir	PSMN	SS
British pine *	PNSY	SS
Strength class C24		
European redwood/whitewood	WPPA	SS
British grown larch	WLAD	SS
Douglas fir (imported)	PSMN	SS
Southern pine (imported)	WPNE	SS
Temperate hardwood:		
Standard name		
European oak	QCXE	TH1

- Machine graded direct to strength class
* British pine SS grade is allocated to strength class C22 but the span tables for C18 can be used.

AII.1 Design loads

The following imposed loads have been considered in the preparation of the span tables. Reference has been made to BS 6399: Part 1: 1996 where appropriate.
Allowance for construction self-weight of up to 0.25 kN/m^2 has been included.

Table AII.2 Imposed loads used in span tables

Application	**Uniformly Distributed Load †† kN/m^2**	**Concentrated Load ††† kN**
Domestic applications (single occupancy)	1.5	0.9
Commercial use - dining areas, cafes, restaurants, balconies †	4.0	1.4

† Suitability of imposed loads for these applications should be verified for each specific project.

†† Used in conjunction with a span/400 deflection limit to enhance dynamic performance and user comfort.

††† BS 6399: Part 1: 1996 outlines concentrated loads appropriate to specific applications for buildings in regular use. However, conventional BS 6399 concentrated loads are not considered appropriate to the design of timber decking intended primarily for leisure activities. As such the 0.9 kN and 1.4 kN concentrated loads have been derived and used in preparation of the deck board span tables. Special account should be taken of any more onerous loading conditions such as large planters, spa pools, ornaments etc.

AII.2 Deck board span tables for double span support condition

Table AII.3 Maximum deck board support centres: Domestic applications

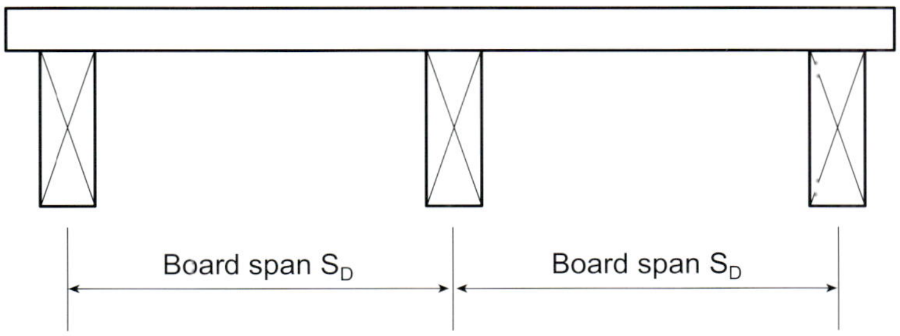

Nominal board size (mm)	**Actual board size** (mm)	**C16**	**C18**	**C24**	**Oak TH1**
		Maximum support centres S_C (mm)			
75 x 25	69 x 22	-	-	300	350
100 x 25	94 x 22	-	300	350	400
125 x 25	119 x 22	350	350	400	450
150 x 25	144 x 22	400	400	450	500
75 x 38	69 x 34	500	550	600	600
100 x 38	94 x 34	600	600	600	600
125 x 38	119 x 34	600	600	600	600
150 x 38	144 x 34	600	600	600	600

Suitable for imposed uniformly distributed loads up to 1.5 kN/m^2.

Suitable for imposed concentrated loads up to 0.9 kN. Treated as medium-term duration.

In calculating span tables, deflection limits of span/400 for UDL and span/250 for concentrated load have been applied.

Tables are calculated assuming smooth boards. Should profiled deck boards be required, additional thickness decking is necessary.

Note: The span tables in this manual have been prepared to assist in the preliminary design of timber decks, eg for species selection and initial member sizing. The detailed structural design for an individual deck will need to be confirmed by an appropriately qualified engineer.

Table AII.4 Maximum deck board support centres: Commercial applications

Nominal board size (mm)	**Actual board size** (mm)	**C16**	**C18**	**C24**	**Oak TH1**
		Maximum support centres S_D (mm)			
75 x 25	69 x 22	-	-	-	-
100 x 25	94 x 22	-	-	-	300
125 x 25	119 x 22	-	-	300	350
150 x 25	144 x 22	-	300	350	400
75 x 38	69 x 34	-	-	450	550
100 x 38	94 x 34	450	450	550	600
125 x 38	119 x 34	550	600	600	600
150 x 38	144 x 34	600	600	600	600

Suitable for imposed uniformly distributed loads up to 4.0 kN/m^2.

Suitable for imposed concentrated loads up to 0.9 kN. Treated as medium-term duration.

In calculating span tables, deflection limits of span/400 for UDL and span/250 for concentrated load have been applied.

Tables are calculated assuming smooth boards. Should profiled deck boards be required, additional thickness decking is necessary.

Note: The span tables in this manual have been prepared to assist in the preliminary design of timber decks, eg for species selection and initial member sizing. The detailed structural design for an individual deck will need to be confirmed by an appropriately qualified engineer.

AII.3 Joist span tables

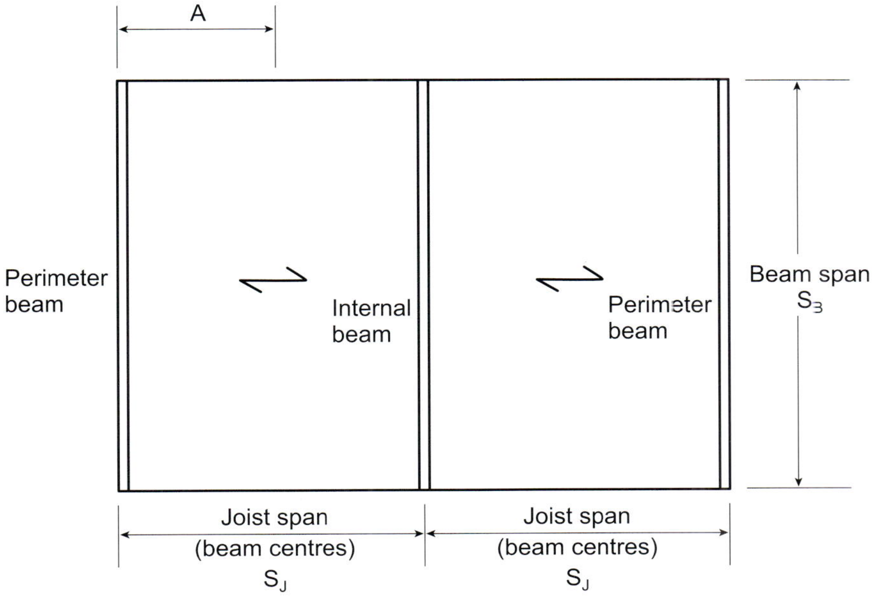

For perimeter beams, A = beam centres = 0.5 joist span

Notes to Joist Span Tables AII.5 - AII.8

Span tables prepared following the principles of BS 5268-7.1: 1989 Structural use of timber – Domestic floor joists.

Floor joist slab loading has not been included.

A dead load of 0.25 kN/m^2 has been included for self-weight of construction.

A deflection limit of span/400 has been applied to limit movement and vibration in use.

Floor joist spans have been limited to a maximum of 4.2 m. Longer spans can be calculated as appropriate. Spans less than 1.0 m have been omitted.

Provide 1 row of strutting to centre of joist spans of 2.5 m or greater.

For 38 x 195 mm and 38 x 220 mm joist sections, provide strutting between joists at intervals not exceeding 6 times joist depth.

Strutting should be blocked to perimeter edge beam and/or wall of adjacent building wherever possible.

Strutting is required where joists span continuously over internal beam supports

Note: The span tables in this manual have been prepared to assist in the <u>preliminary</u> design of timber decks, eg for species selection and initial member sizing. The detailed structural design for an individual deck will need to be confirmed by an appropriately qualified engineer.

Table AII.5 Maximum joist spans - Strength Class C16

Joist section (mm)		Imposed load 1.5 kN/m^2			Imposed load 4.0 kN/m^2		
		Joist centres (mm)			Joist centres (mm)		
		400	500	600	400	500	600
width	depth	Maximum clear joist span S_J (m)					
38 x	97	1.86	1.73	1.63	1.34	1.20	1.09
	122	2.34	2.17	2.04	1.66	1.49	1.36
	147	2.82	2.61	2.46	1.98	1.77	1.62
	170	3.25	3.02	2.84	2.27	2.03	1.86
	195	3.72	3.46	3.25	2.58	2.31	2.11
	220	4.19	3.89	3.66	2.89	2.59	2.37
47 x	97	2.00	1.86	1.75	1.47	1.33	1.22
	122	2.51	2.33	2.20	1.85	1.66	1.51
	147	3.02	2.81	2.64	2.20	1.97	1.80
	170	3.49	3.24	3.05	2.52	2.26	2.07
	195	4.00	3.71	3.50	2.87	2.57	2.35
	220	4.20	4.18	3.94	3.21	2.88	2.63
63 x	122	2.77	2.58	2.42	2.05	1.90	1.75
	147	3.33	3.10	2.92	2.47	2.29	2.09
	170	3.84	3.57	3.37	2.85	2.62	2.39
	195	4.20	4.09	3.85	3.27	2.98	2.72
	220	4.20	4.20	4.20	3.69	3.33	3.05
	245	4.20	4.20	4.20	4.10	3.68	3.37
75 x	122	2.93	2.73	2.57	2.18	2.02	1.90
	147	3.53	3.28	3.09	2.62	2.43	2.28
	170	4.06	3.78	3.57	3.03	2.81	2.61
	195	4.20	4.20	4.08	3.47	3.22	2.97
	220	4.20	4.20	4.20	3.91	3.63	3.32
	245	4.20	4.20	4.20	4.20	4.02	3.67

Note: The span tables in this manual have been prepared to assist in the preliminary design of timber decks, eg for species selection and initial member sizing. The detailed structural design for an individual deck will need to be confirmed by an appropriately qualified engineer.

See also Notes to Joist span tables on page 61.

Table AII.6 Maximum joist spans - Strength Class C18

Joist section (mm)		Imposed load 1.5 kN/m²			Imposed load 4.0 kN/m²		
		Joist centres (mm)			Joist centres (mm)		
		400	500	600	400	500	600
width	depth	Maximum clear joist span S_J (m)					
38 x	97	1.89	1.75	1.64	1.38	1.25	1.15
	122	2.37	2.20	2.06	1.74	1.56	1.42
	147	2.85	2.64	2.49	2.07	1.85	1.69
	170	3.29	3.05	2.87	2.37	2.13	1.94
	195	3.76	3.50	3.29	2.70	2.42	2.21
	220	4.20	3.94	3.71	3.02	2.71	2.47
47 x	97	2.03	1.88	1.77	1.49	1.38	1.28
	122	2.54	2.36	2.22	1.87	1.73	1.58
	147	3.06	2.84	2.67	2.26	2.06	1.89
	170	3.53	3.28	3.09	2.61	2.37	2.16
	195	4.04	3.75	3.53	2.99	2.69	2.46
	220	4.20	4.20	3.98	3.36	3.01	2.75
63 x	122	2.80	2.60	2.45	2.08	1.92	1.81
	147	3.37	3.13	2.95	2.50	2.32	2.18
	170	3.88	3.61	3.40	2.89	2.68	2.50
	195	4.20	4.14	3.90	3.31	3.07	2.85
	220	4.20	4.20	4.20	3.73	3.46	3.19
	245	4.20	4.20	4.20	4.15	3.85	3.53
75 x	122	2.97	2.76	2.60	2.21	2.05	1.92
	147	3.56	3.32	3.12	2.65	2.46	2.31
	170	4.11	3.83	3.61	3.06	2.84	2.67
	195	4.20	4.20	4.13	3.51	3.26	3.06
	220	4.20	4.20	4.20	3.96	3.67	3.45
	245	4.20	4.20	4.20	4.20	4.08	3.84

Note: The span tables in this manual have been prepared to assist in the preliminary design of timber decks, eg for species selection and initial member sizing. The detailed structural design for an individual deck will need to be confirmed by an appropriately qualified engineer.

See also Notes to Joist span tables on page 61.

Table AII.7 Maximum joist spans - Strength Class C24

Joist section (mm)		Imposed load 1.5 kN/m^2			Imposed load 4.0 kN/m^2		
		Joist centres (mm)			Joist centres (mm)		
		400	500	600	400	500	600
width	depth	Maximum clear joist span S_J (m)					
38 x	97	2.00	1.85	1.74	1.47	1.36	1.28
	122	2.51	2.33	2.19	1.85	1.71	1.60
	147	3.02	2.80	2.63	2.22	2.06	1.93
	170	3.48	3.23	3.04	2.57	2.38	2.21
	195	3.99	3.70	3.49	2.94	2.73	2.51
	220	4.20	4.17	3.93	3.32	3.07	2.81
47 x	97	2.15	1.99	1.87	1.58	1.47	1.38
	122	2.69	2.50	2.35	1.99	1.84	1.73
	147	3.24	3.01	2.83	2.40	2.22	2.08
	170	3.74	3.47	3.27	2.77	2.56	2.41
	195	4.20	3.98	3.75	3.17	2.94	2.76
	220	4.20	4.20	4.20	3.57	3.31	3.11
63 x	122	2.97	2.76	2.60	2.20	2.04	1.92
	147	3.56	3.32	3.12	2.65	2.46	2.31
	170	4.11	3.83	3.60	3.06	2.84	2.67
	195	4.20	4.20	4.13	3.51	3.25	3.06
	220	4.20	4.20	4.20	3.95	3.67	3.45
	245	4.20	4.20	4.20	4.20	4.08	3.84
75 x	122	3.14	2.92	2.75	2.34	2.17	2.04
	147	3.77	3.51	3.31	2.81	2.61	2.45
	170	4.35	4.05	3.82	3.25	3.02	2.84
	195	4.20	4.20	4.20	3.72	3.45	3.25
	220	4.20	4.20	4.20	4.19	3.89	3.66
	245	4.20	4.20	4.20	4.20	4.20	4.07

Note: The span tables in this manual have been prepared to assist in the preliminary design of timber decks, eg for species selection and initial member sizing. The detailed structural design for an individual deck will need to be confirmed by an appropriately qualified engineer.

See also Notes to Joist span tables on page 61.

Table AII.8 Maximum joist spans - Oak TH1

Joist section (mm)		Imposed load 1.5 kN/m²			Imposed load 4.0 kN/m²		
		Joist centres (mm)			Joist centres (mm)		
		400	500	600	400	500	600
width	depth	Maximum clear joist span S_J (m)					
38 x	97	2.10	1.95	1.83	1.55	1.44	1.35
	122	2.63	2.44	2.30	1.95	1.80	1.69
	147	3.16	2.94	2.77	2.34	2.17	2.04
	170	3.64	3.39	3.19	2.71	2.51	2.36
	195	4.17	3.88	3.65	3.10	2.87	2.70
47 x	97	2.25	2.09	1.97	1.67	1.55	1.45
	122	2.82	2.62	2.47	2.09	1.94	1.82
	147	3.38	3.15	2.97	2.52	2.34	2.20
	170	3.90	3.63	3.42	2.91	2.70	2.54
	195	4.20	4.16	3.92	3.33	3.09	2.91
	220	4.20	4.20	4.20	3.76	3.49	3.28
63 x	122	3.10	2.88	2.72	2.31	2.15	2.02
	147	3.72	3.46	3.27	2.78	2.58	2.43
	170	4.20	3.99	3.77	3.21	2.98	2.81
	195	4.20	4.20	4.20	3.68	3.42	3.22
	220	4.20	4.20	4.20	4.14	3.85	3.62
	245	4.20	4.20	4.20	4.20	4.20	4.20
75 x	122	3.27	3.05	2.88	2.45	2.28	2.14
	147	3.92	3.66	3.45	2.95	2.74	2.58
	170	4.20	4.20	3.98	3.40	3.16	2.98
	195	4.20	4.20	4.20	3.90	3.62	3.41
	220	4.20	4.20	4.20	4.20	4.08	3.84
	245	4.20	4.20	4.20	4.20	4.20	4.20

Note: The span tables in this manual have been prepared to assist in the <u>preliminary</u> design of timber decks, eg for species selection and initial member sizing. The detailed structural design for an individual deck will need to be confirmed by an appropriately qualified engineer.

See also Notes to Joist span tables on page 61.

AII.4 Beam span tables

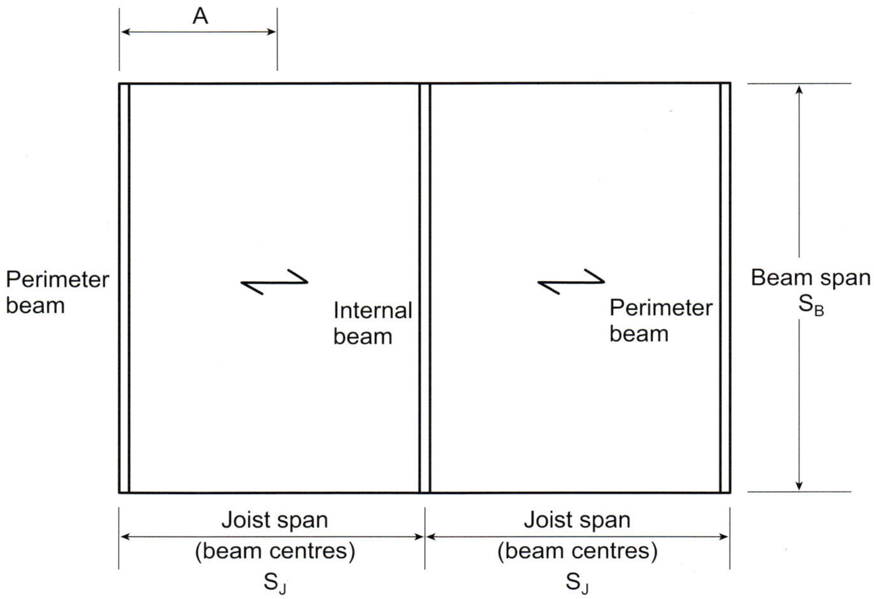

For perimeter beams, A = beam centres = 0.5 joist span

Notes to Beam Span Tables AII.9 to AII.20

Span tables prepared following the principles of BS 5268-7.

Span have been limited to a maximum of 4.2 m. Longer spans can be calculated as appropriate. Spans less than 1.0 m have been omitted.

A deflection limit of span/400 has been applied to limit movement and vibration in use.

A dead load of 0.25 kN/m^2 has been included to cater for self-weight of construction.

Double member beams should be securely fixed together to ensure composite action.

Where spaced double member beams are used to support joists from either side at the same level, the beam members should be adequately fastened together at each joist support position.

Where spaced double member beams are used to support continuous joists spanning-over, adequate fixing between the beam members is recommended at each joist position. This may be omitted where the beam is loaded equally from each side by joists of equal span.

Notching and drilling of beams is not permitted unless calculated in accordance with BS 5268- 2: 2002.

Table AII.9 Single member beams - Strength Class C16

Beam size (mm)		Imposed load 1.5 kN/m^2					
		Beam centres (m)					
		1.2	1.8	2.4	3.0	3.6	4.2
width	depth	Maximum clear joist span S_B (m)					
47 x	122	1.48	1.27	1.11	-	-	-
	147	1.78	1.53	1.32	-	-	-
	170	2.06	1.76	1.51	1.31	1.08	-
	195	2.36	2.00	1.72	1.50	1.24	1.06
	220	2.67	2.24	1.93	1.69	1.40	1.19
63 x	122	1.65	1.42	1.28	1.15	1.04	-
	147	1.98	1.71	1.54	1.37	1.24	1.08
	170	2.29	1.98	1.76	1.57	1.43	1.25
	195	2.63	2.27	2.00	1.79	1.62	1.43
	220	2.96	2.56	2.24	2.00	1.82	1.62
	245	3.29	2.85	2.48	2.21	2.01	1.80
75 x	122	1.75	1.52	1.36	1.25	1.14	1.05
	147	2.11	1.83	1.64	1.50	1.36	1.26
	170	2.44	2.11	1.90	1.72	1.56	1.44
	195	2.79	2.42	2.18	1.95	1.78	1.64
	220	3.15	2.73	2.45	2.19	1.99	1.84
	245	3.50	3.03	2.71	2.42	2.20	2.03

Note: The span tables in this manual have been prepared to assist in the preliminary design of timber decks, eg for species selection and initial member sizing. The detailed structural design for an individual deck will need to be confirmed by an appropriately qualified engineer.

See also Notes to Beam span tables on page 66.

Table AII.10 Single member beams - Strength Class C18

Beam size (mm)		Imposed load 1.5 kN/m^2					
		Beam centres (m)					
		1.2	1.8	2.4	3.0	3.6	4.2
width	depth	Maximum clear joist span S_B (m)					
47 x	122	1.50	1.29	1.16	-	-	-
	147	1.81	1.55	1.38	1.13	-	-
	170	2.09	1.80	1.58	1.31	1.08	-
	195	2.39	2.06	1.80	1.50	1.24	1.06
	220	2.70	2.32	2.02	1.69	1.40	1.19
63 x	122	1.67	1.44	1.29	1.19	1.05	-
	147	2.01	1.73	1.56	1.43	1.27	1.08
	170	2.32	2.00	1.80	1.64	1.46	1.25
	195	2.66	2.30	2.06	1.87	1.68	1.43
	220	2.99	2.59	2.33	2.09	1.89	1.62
	245	3.33	2.88	2.59	2.31	2.10	1.80
75 x	122	1.77	1.53	1.38	1.27	1.18	1.07
	147	2.13	1.85	1.66	1.53	1.43	1.29
	170	2.47	2.13	1.92	1.77	1.63	1.50
	195	2.82	2.45	2.20	2.03	1.86	1.71
	220	3.18	2.76	2.48	2.28	2.08	1.92
	245	3.54	3.07	2.76	2.53	2.31	2.13

Note: The span tables in this manual have been prepared to assist in the preliminary design of timber decks, eg for species selection and initial member sizing. The detailed structural design for an individual deck will need to be confirmed by an appropriately qualified engineer.

See also Notes to Beam span tables on page 66.

Table AII.11 Single member beams - Strength Class C24

Beam size (mm)		Imposed load 1.5 kN/m²					
		Beam centres (m)					
		1.2	1.8	2.4	3.0	3.6	4.2
width	depth	Maximum clear joist span S_B (m)					
47 x	122	1.60	1.38	1.24	1.00	-	-
	147	1.93	1.66	1.49	1.20	1.00	-
	170	2.23	1.92	1.72	1.39	1.15	-
	195	2.55	2.20	1.98	1.59	1.32	1.13
	220	2.88	2.48	2.23	1.79	1.49	1.27
63 x	122	1.78	1.54	1.38	1.27	1.12	-
	147	2.14	1.85	1.67	1.53	1.35	1.15
	170	2.47	2.14	1.92	1.77	1.56	1.33
	195	2.80	2.45	2.21	2.03	1.78	1.52
	220	3.19	2.76	2.49	2.29	2.01	1.72
	245	3.55	3.08	2.77	2.55	2.24	1.91
75 x	122	1.87	1.64	1.47	1.36	1.27	1.14
	147	2.27	1.97	1.78	1.64	1.53	1.37
	170	2.63	2.28	2.05	1.89	1.76	1.59
	195	3.01	2.61	2.35	2.17	2.02	1.82
	220	3.39	2.94	2.65	2.44	2.28	2.05
	245	3.77	3.27	2.95	2.72	2.54	2.28

Note: The span tables in this manual have been prepared to assist in the <u>preliminary</u> design of timber decks, eg for species selection and initial member sizing. The detailed structural design for an individual deck will need to be confirmed by an appropriately qualified engineer.

See also Notes to Beam span tables on page 66.

Table AII.12 Single member beams - Oak TH1

Beam size (mm)		Imposed load 1.5 kN/m^2					
		Beam centres (m)					
		1.2	1.8	2.4	3.0	3.6	4.2
width	depth	Maximum clear joist span S_B (m)					
47 x	122	1.70	1.47	1.32	1.22	1.14	1.07
	147	2.05	1.77	1.60	1.47	1.37	1.29
	170	2.37	2.05	1.84	1.70	1.58	1.49
	195	2.71	2.35	2.11	1.94	1.81	1.71
	220	3.05	2.65	2.38	2.19	2.04	1.92
63 x	122	1.88	1.64	1.48	1.36	1.27	1.20
	147	2.27	1.97	1.78	1.64	1.53	1.44
	170	2.62	2.27	2.05	1.89	1.77	1.67
	195	3.00	2.61	2.35	2.17	2.03	1.91
	220	3.38	2.94	2.65	2.44	2.28	2.16
	245	3.75	3.27	2.95	2.72	2.54	2.40
75 x	122	2.00	1.74	1.57	1.45	1.36	1.28
	147	2.41	2.09	1.89	1.74	1.63	1.54
	170	2.78	2.42	2.18	2.02	1.89	1.78
	195	3.18	2.77	2.50	2.31	2.16	2.04
	220	3.58	3.12	2.82	2.61	2.44	2.30
	245	3.98	3.47	3.14	2.90	2.71	2.56

Note: The span tables in this manual have been prepared to assist in the preliminary design of timber decks, eg for species selection and initial member sizing. The detailed structural design for an individual deck will need to be confirmed by an appropriately qualified engineer.

See also Notes to Beam span tables on page 66.

Table AII.13 Double member beams - Strength Class C16

Beam size (mm)		Imposed load 1.5 kN/m^2					
		Beam centres (m)					
		1.2	1.8	2.4	3.0	3.6	4.2
width	depth	Maximum clear joist span S_B (m)					
2 x 38 x	122	1.84	1.60	1.44	1.32	1.21	1.12
	147	2.22	1.92	1.73	1.59	1.44	1.33
	170	2.56	2.22	2.00	1.82	1.65	1.53
	195	2.94	2.55	2.30	2.07	1.88	1.74
	220	3.31	2.87	2.59	2.32	2.11	1.95
2 x 47 x	122	1.99	1.72	1.55	1.43	1.34	1.25
	147	2.39	2.07	1.87	1.73	1.61	1.49
	170	2.76	2.40	2.16	1.99	1.85	1.7[illegible]
	195	3.16	2.75	2.48	2.29	2.10	1.94
	220	3.56	3.10	2.80	2.58	2.35	2.17
2 x 63 x	122	2.19	1.91	1.73	1.59	1.49	1.41
	147	2.64	2.30	2.08	1.92	1.80	1.70
	170	3.05	2.65	2.40	2.22	2.08	1.96
	195	3.49	3.04	2.75	2.54	2.38	2.25
	220	3.93	3.43	3.10	2.87	2.68	2.53
	245	4.20	3.81	3.45	3.19	2.99	2.79
2 x 75 x	122	2.33	2.03	1.84	1.70	1.59	1.50
	147	2.80	2.44	2.21	2.04	1.91	1.81
	170	3.23	2.82	2.55	2.36	2.21	2.09
	195	3.70	3.23	2.92	2.70	2.54	2.40
	220	4.16	3.64	3.30	3.05	2.86	2.70
	245	4.20	4.04	3.67	3.39	3.18	3.01

Note: The span tables in this manual have been prepared to assist in the preliminary design of timber decks, eg for species selection and initial member sizing. The detailed structural design for an individual deck will need to be confirmed by an appropriately qualified engineer.

See also Notes to Beam span tables on page 66.

Table AII.14 Double member beams - Strength Class C18

Beam size (mm)		Imposed load 1.5 kN/m^2					
		Beam centres (m)					
		1.2	1.8	2.4	3.0	3.6	4.2
width	depth	Maximum clear joist span S_B (m)					
2 x 38 x	122	1.87	1.62	1.46	1.34	1.25	1.17
	147	2.25	1.95	1.75	1.61	1.51	1.39
	170	2.59	2.25	2.03	1.87	1.73	1.60
	195	2.97	2.58	2.32	2.14	1.97	1.82
	220	3.35	2.91	2.62	2.41	2.21	2.04
2 x 47 x	122	2.01	1.74	1.57	1.45	1.36	1.28
	147	2.42	2.10	1.89	1.75	1.63	1.54
	170	2.79	2.43	2.19	2.02	1.89	1.78
	195	3.20	2.78	2.51	2.31	2.16	2.03
	220	3.60	3.13	2.83	2.61	2.44	2.27
2 x 63 x	122	2.22	1.93	1.75	1.61	1.51	1.43
	147	2.67	2.33	2.10	1.94	1.82	1.72
	170	3.08	2.69	2.43	2.24	2.10	1.99
	195	3.53	3.08	2.78	2.57	2.41	2.28
	220	3.97	3.47	3.14	2.90	2.72	2.57
	245	4.20	3.86	3.49	3.23	3.02	2.86
2 x 75 x	122	2.35	2.05	1.86	1.72	1.61	1.52
	147	2.83	2.47	2.24	2.07	1.94	1.83
	170	3.27	2.85	2.58	2.39	2.24	2.12
	195	3.74	3.27	2.96	2.74	2.57	2.43
	220	4.20	3.68	3.33	3.08	2.89	2.74
	245	4.20	4.09	3.71	3.43	3.22	3.05

Note: The span tables in this manual have been prepared to assist in the preliminary design of timber decks, eg for species selection and initial member sizing. The detailed structural design for an individual deck will need to be confirmed by an appropriately qualified engineer.

See also Notes to Beam span tables on page 66.

Table AII.15 Double member beams - Strength Class C24

Beam size (mm)		Imposed load 1.5 kN/m^2					
		Beam centres (m)					
		1.2	**1.8**	**2.4**	**3.0**	**3.6**	**4.2**
width	**depth**	**Maximum clear joist span S_B (m)**					
2 x 38 x	122	1.99	1.72	1.55	1.43	1.34	1.26
	147	2.39	2.08	1.87	1.73	1.61	1.52
	170	2.76	2.40	2.16	1.99	1.86	1.76
	195	3.16	2.75	2.48	2.29	2.14	2.0[illegible]
	220	3.57	3.10	2.80	2.58	2.41	2.27
2 x 47 x	122	2.14	1.86	1.68	1.55	1.45	1.37
	147	2.57	2.24	2.02	1.86	1.74	1.65
	170	2.97	2.58	2.34	2.16	2.02	1.90
	195	3.40	2.96	2.68	2.47	2.31	2.18
	220	3.83	3.34	3.02	2.79	2.61	2.46
2 x 63 x	122	2.36	2.06	1.86	1.72	1.61	1.52
	147	2.84	2.48	2.24	2.07	1.94	1.84
	170	3.28	2.86	2.59	2.39	2.24	2.12
	195	3.75	3.27	2.97	2.74	2.57	2.43
	220	4.22	3.69	3.34	3.09	3.09	2.74
	245	4.20	4.10	3.72	3.44	3.23	3.05
2 x 75 x	122	2.50	2.18	1.98	1.83	1.72	1.62
	147	3.01	2.63	2.38	2.20	2.07	1.96
	170	3.47	3.03	2.75	2.55	2.39	2.26
	195	3.97	3.47	3.15	2.92	2.74	2.59
	220	4.20	3.91	3.55	3.29	3.08	2.92
	245	4.20	4.20	3.95	3.66	3.43	3.25

Note: The span tables in this manual have been prepared to assist in the <u>preliminary</u> design of timber decks, eg for species selection and initial member sizing. The detailed structural design for an individual deck will need to be confirmed by an appropriately qualified engineer.

See also Notes to Beam span tables on page 66.

Table AII.16 Double member beams - Oak TH1

Beam size (mm)		Imposed load 1.5 kN/m^2					
		Beam centres (m)					
		1.2	1.8	2.4	3.0	3.6	4.2
width	depth	Maximum clear joist span S_B (m)					
2 x 38 x	122	2.05	1.78	1.61	1.49	1.39	1.32
	147	2.47	2.15	1.94	1.79	1.68	1.58
	170	2.85	2.48	2.24	2.07	1.94	1.83
	195	3.26	2.84	2.57	2.37	2.22	2.10
	220	3.67	3.20	2.90	2.68	2.50	2.37
2 x 47 x	122	2.20	1.92	1.74	1.61	1.50	1.42
	147	2.65	2.31	2.09	1.93	1.81	1.71
	170	3.06	2.67	2.42	2.23	2.09	1.98
	195	3.50	3.06	2.77	2.56	2.40	2.27
	220	3.94	3.44	3.12	2.89	2.70	2.56
2 x 63 x	122	2.43	2.12	1.92	1.78	1.67	1.58
	147	2.92	2.55	2.31	2.14	2.01	1.90
	170	3.37	2.95	2.67	2.48	2.32	2.20
	195	3.85	3.37	3.06	2.84	2.66	2.52
	220	4.20	3.80	3.45	3.20	3.00	2.84
	245	4.20	4.20	3.83	3.55	3.34	3.16
2 x 75 x	122	2.57	2.25	2.04	1.89	1.78	1.68
	147	3.09	2.70	2.46	2.28	2.14	2.02
	170	3.56	3.12	2.84	2.63	2.47	2.34
	195	4.07	3.57	3.25	3.01	2.83	2.68
	220	4.20	4.02	3.66	3.39	3.19	3.02
	245	4.20	4.20	4.06	3.77	3.54	3.36

Note: The span tables in this manual have been prepared to assist in the preliminary design of timber decks, eg for species selection and initial member sizing. The detailed structural design for an individual deck will need to be confirmed by an appropriately qualified engineer.

See also Notes to Beam span tables on page 66.

Table AII.17 Double member beams - Strength Class C16

Beam size (mm)		Imposed load 4.0 kN/m^2					
		Beam centres (m)					
		1.2	1.8	2.4	3.0	3.6	4.2
width	depth	Maximum clear joist span S_B (m)					
2 x 38 x	122	1.34	1.09	-	-	-	-
	147	1.61	1.30	1.03	-	-	-
	170	1.85	1.50	1.19	-	-	-
	195	2.10	1.70	1.37	1.09	-	-
	220	2.35	1.91	1.54	1.22	1.01	-
2 x 47 x	122	1.45	1.22	1.05	-	-	-
	147	1.74	1.46	1.25	1.02	-	-
	170	2.02	1.67	1.44	1.18	-	-
	195	2.31	1.90	1.64	1.35	1.12	-
	220	2.61	2.13	1.83	1.53	1.26	1.08
2 x 63 x	122	1.61	1.39	1.23	1.09	-	-
	147	1.94	1.67	1.46	1.30	1.15	-
	170	2.24	1.93	1.67	1.49	1.32	1.13
	195	2.57	2.21	1.91	1.70	1.52	1.29
	220	2.90	2.47	2.13	1.90	1.71	1.46
	245	3.22	2.74	2.36	2.10	1.90	1.62
2 x 75 x	122	1.71	1.48	1.33	1.19	1.09	-
	147	2.06	1.78	1.60	1.42	1.29	1.17
	170	2.38	2.06	1.83	1.63	1.48	1.35
	195	2.73	2.36	2.08	1.86	1.69	1.55
	220	3.08	2.66	2.33	2.08	1.89	1.74
	245	3.43	2.97	2.58	2.30	2.09	1.93

Note: The span tables in this manual have been prepared to assist in the <u>preliminary</u> design of timber decks, eg for species selection and initial member sizing. The detailed structural design for an individual deck will need to be confirmed by an appropriately qualified engineer.

See also Notes to Beam span tables on page 66.

Table AII.18 Double member beams - Strength Class C18

Beam size (mm)		Imposed load 4.0 kN/m^2					
		Beam centres (m)					
		1.2	1.8	2.4	3.0	3.6	4.2
width	depth	Maximum clear joist span S_B (m)					
2 x 38 x	122	1.35	1.14	-	-	-	-
	147	1.63	1.36	1.03	-	-	-
	170	1.89	1.56	1.19	-	-	-
	195	2.16	1.78	1.37	1.09	-	-
	220	2.44	1.99	1.54	1.22	1.01	-
2 x 47 x	122	1.47	1.26	1.07	-	-	-
	147	1.77	1.52	1.29	1.02	-	-
	170	2.04	1.75	1.49	1.18	-	-
	195	2.34	1.99	1.70	1.35	1.12	-
	220	2.64	2.23	1.92	1.53	1.26	1.08
2 x 63 x	122	1.63	1.41	1.26	1.14	-	-
	147	1.86	1.69	1.52	1.36	1.15	-
	170	2.27	1.96	1.75	1.56	1.32	1.13
	195	2.60	2.24	1.99	1.77	1.52	1.29
	220	2.93	2.53	2.23	1.99	1.71	1.46
	245	3.26	2.82	2.47	2.20	1.90	1.62
2 x 75 x	122	1.73	1.50	1.35	1.24	1.14	-
	147	2.09	1.81	1.62	1.49	1.35	1.17
	170	2.41	2.09	1.88	1.71	1.55	1.35
	195	2.76	2.39	2.15	1.84	1.77	1.55
	220	3.12	2.70	2.43	2.18	1.98	1.75
	245	3.47	3.00	2.70	2.41	2.19	1.94

Note: The span tables in this manual have been prepared to assist in the preliminary design of timber decks, eg for species selection and initial member sizing. The detailed structural design for an individual deck will need to be confirmed by an appropriately qualified engineer.

See also Notes to Beam span tables on page 66.

Table A.19 Double member beams - Strength Class C24

Beam size (mm)		Imposed load 4.0 kN/m^2					
		Beam centres (m)					
		1.2	1.8	2.4	3.0	3.6	4.2
width	depth	Maximum clear joist span S_B (m)					
2 x 38 x	122	1.45	1.23	-	-	-	-
	147	1.74	1.48	1.10	-	-	-
	170	2.02	1.70	1.27	1.01	-	-
	195	2.31	1.95	1.46	1.16	-	-
	220	2.61	2.20	1.64	1.30	1.08	-
2 x 47 x	122	1.57	1.35	1.13	-	-	-
	147	1.88	1.62	1.37	1.09	-	-
	170	2.18	1.88	1.58	1.26	1.04	-
	195	2.50	2.15	1.81	1.44	1.19	1.02
	220	2.81	2.43	2.04	1.62	1.34	1.15
2 x 63 x	122	1.74	1.50	1.35	1.22	1.01	-
	147	2.09	1.81	1.63	1.47	1.22	1.04
	170	2.42	2.09	1.88	1.70	1.41	1.20
	195	2.77	2.40	2.15	1.94	1.61	1.38
	220	3.12	2.70	2.43	2.19	1.82	1.55
	245	3.48	3.01	2.70	2.44	2.02	1.73
2 x 75 x	122	1.85	1.60	1.44	1.33	1.21	1.03
	147	2.23	1.93	1.74	1.60	1.46	1.24
	170	2.57	2.23	2.01	1.85	1.68	1.44
	195	2.95	2.55	2.30	2.12	1.93	1.65
	220	3.32	2.88	2.59	2.39	2.17	1.86
	245	3.69	3.20	2.89	2.66	2.42	2.06

Note: The span tables in this manual have been prepared to assist in the <u>preliminary</u> design of timber decks, eg for species selection and initial member sizing. The detailed structural design for an individual deck will need to be confirmed by an appropriately qualified engineer.

See also Notes to Beam span tables on page 66.

Table AII.20 Double member beams - Oak TH1

Beam size (mm)		**Imposed load 4.0 kN/m^2**					
		Beam centres (m)					
		1.2	**1.8**	**2.4**	**3.0**	**3.6**	**4.2**
width	**depth**	**Maximum clear joist span S_B (m)**					
2 x 38 x	122	1.50	1.30	1.16	1.06	-	-
	147	1.81	1.56	1.40	1.28	1.19	1.12
	170	2.09	1.80	1.62	1.48	1.38	1.29
	195	2.40	2.07	1.85	1.70	1.58	1.48
	220	2.70	2.33	2.09	1.92	1.78	1.67
2 x 47 x	122	1.62	1.40	1.26	1.16	1.08	1.01
	147	1.95	1.69	1.52	1.39	1.30	1.22
	170	2.26	1.95	1.75	1.61	1.50	1.41
	195	2.59	2.24	2.01	1.85	1.72	1.62
	220	2.91	2.52	2.27	2.08	1.94	1.82
2 x 63 x	122	1.80	1.56	1.40	1.29	1.21	1.14
	147	2.16	1.88	1.69	1.56	1.45	1.37
	170	2.50	2.17	1.95	1.80	1.68	1.58
	195	2.86	2.49	2.24	2.06	1.93	1.82
	220	3.23	2.80	2.53	2.33	2.17	2.05
	245	3.59	3.12	2.81	2.59	2.42	2.28
2 x 75 x	122	1.91	1.66	1.50	1.38	1.29	1.22
	147	2.30	2.00	1.80	1.66	1.55	1.47
	170	2.65	2.31	2.08	1.92	1.79	1.69
	195	3.04	2.64	2.39	2.20	2.06	1.94
	220	3.42	2.98	2.69	2.48	2.32	2.19
	245	3.81	3.31	2.99	2.76	2.58	2.44

Note: The span tables in this manual have been prepared to assist in the preliminary design of timber decks, eg for species selection and initial member sizing. The detailed structural design for an individual deck will need to be confirmed by an appropriately qualified engineer.

See also Notes to Beam span tables on page 66.

Other publications from TRADA Technology

Visit the TRADA Bookshop at www.trada.co.uk:

- for over 200 publications on timber and construction
- to register for regular updates and offers on new publications added to the Bookshop

- TRADA members save up to 50% on TRADA Technology publications; 10% discount on British Standards
- For membership details go to www.trada.co.uk/membership or call 01494 569603 for details